BUILDING INNOVATION
CAPABILITY IN ORGANIZATIONS

An International Cross-Case Perspective

T0339441

Series on Technology Management*

Series Editor: J. Tidd (Univ. of Sussex, UK)   ISSN 0219-9823

Published

Vol. 3   From Knowledge Management to Strategic Competence
Measuring Technological, Market and Organisational Innovation (2nd Edition)
edited by J. Tidd *(Univ. of Sussex, UK)*

Vol. 4   Japanese Cost Management
edited by Y. Monden *(Univ. of Tsukuba, Japan)*

Vol. 5   R&D Strategy on Organisation
Managing Technical Change in Dynamic Contexts
by V. Chiesa *(Univ. degli Studi di Milano, Italy)*

Vol. 6   Social Interaction and Organisational Change
Aston Perspectives on Innovation Networks
edited by O. Jones *(Aston Univ., UK)*, S. Conway *(Aston Univ., UK)*
& F. Steward *(Aston Univ., UK)*

Vol. 7   Innovation Management in the Knowledge Economy
edited by B. Dankbaar *(Univ. of Nijmegen, The Netherlands)*

Vol. 8   Digital Innovation
Innovation Processes in Virtual Clusters and Digital Regions
edited by G. Passiante *(Univ. of Lecce, Italy)*, V. Elia *(Univ. of Lecce, Italy)* & T. Massari *(Univ. of Lecce, Italy)*

Vol. 9   Service Innovation
Organisational Responses to Technological Opportunities and Market Imperatives
edited by J. Tidd *(Univ. of Sussex, UK)* & F. M. Hull *(Fordham Univ., USA)*

Vol. 10   Open Source
A Multidisciplinary Approach
by M. Muffatto *(University of Padua, Italy)*

Vol. 11   Involving Customers in New Service Development
edited by B. Edvardsson, A. Gustafsson, P. Kristensson,
P. Magnusson & J. Matthing *(Karlstad University, Sweden)*

Vol. 12   Project-Based Organization in the Knowledge-Based Society
by M. Kodama *(Nihon University, Japan)*

Vol. 13   Building Innovation Capability in Organizations
An International Cross-Case Perspective
by M. Terziovski *(University of Melbourne, Australia)*

*For the complete list of titles in this series, please write to the Publisher.

SERIES ON TECHNOLOGY MANAGEMENT – VOL. 13

BUILDING INNOVATION CAPABILITY IN ORGANIZATIONS

An International Cross-Case Perspective

Milé Terziovski

University of Melbourne, Australia

Imperial College Press

ICP

Published by

Imperial College Press
57 Shelton Street
Covent Garden
London WC2H 9HE

Distributed by

World Scientific Publishing Co. Pte. Ltd.
5 Toh Tuck Link, Singapore 596224
USA office: 27 Warren Street, Suite 401-402, Hackensack, NJ 07601
UK office: 57 Shelton Street, Covent Garden, London WC2H 9HE

Library of Congress Cataloging-in-Publication Data
Terziovski, Milé.
    Building innovation capability in organizations : an international cross-case
perspective / by Milé Terziovski.
        p. cm. -- (Series on technology management ; vol. 13)
    Includes bibliographical references and index.
    ISBN-13: 978-1-86094-765-0
    ISBN-10: 1-86094-765-4
    1. Technological innovations--Case studies. 2. International business
enterprises--Case studies. 3. New products--Case studies. 4. Electronic
commerce--Case studies. 5. Sustainable development--Case studies. I. Title.
    HD45.T3985 2007
    658.4'063--dc22

                                                        2007029986

British Library Cataloguing-in-Publication Data
A catalogue record for this book is available from the British Library.

Typeset by Stallion Press
Email: enquiries@stallionpress.com

Printed in Singapore by World Scientific Printers (S) Pte Ltd

*To my family whose unconditional love and faith
has been a source of inspiration*

Preface

As organizations integrate further into the global economy, managers need to predict their competitive outlook and rapidly adapt their innovation strategies to meet their competitors head on. Many studies have been published and indeed special issues of top flight management journals have been devoted to innovation, yet the drivers of success in systematic innovation are not well understood from a theoretical level, hence existing practical insights are not well-founded.

The book takes an international cross-case perspective to explore how organizations build innovation capability. Eight case studies were developed in France, Germany, Australia and Thailand, in various industry sectors, ranging from manufacturing, mining, biotechnology and computers. Each case study explores the role of the central constructs of e-Commerce, sustainable development orientation (SDO) and New Product Development (NPD) in developing innovation capability for the respective organization.

The book justifies the efficacy of building innovation capability so that the present manager's perception of innovation as a technically-driven strategy is expanded to include innovation as a competitive business strategy. The book primarily informs managers and management researchers in terms of "what works, why and how it works" in building innovation-driven organizations. Based on qualitative multiple cross-case analysis, the book articulates the dimensions of innovation capability in organizations, and how innovation capability is developed and exploited. The key drivers of innovative organizations are also identified.

The "breakthrough" contribution of this book is to identify and crisply articulate innovation capability in a comprehensive manner, and as an integrative construct. Most previous books have focused on only one aspect of innovation capability, perhaps reflecting the specialized nature of these past researchers. This book takes an integrative multi-case study perspective.

The book develops new theory in the field of innovation management that leads to new conclusions of general value and practical insights for executives wishing to improve the innovation outcomes of their firms. Conversely, we identify which practices create the greatest barriers to innovation, and hence should be addressed in more detail during the commercialization process. This will prevent the waste of scarce organizational resources. Identifying time delays between the introduction of an innovative idea and improvement in the company's corporate performance is valuable because it will help reassure managers that innovation is a worthwhile investment, even if it does take time.

Milé Terziovski
Melbourne, Australia
February 2007

Acknowledgments

I wish to acknowledge the Australian Research Council (ARC) for providing funds for the development of the case studies and my co-grantee, Professor Danny Samson, who has had a significant impact on my academic career.

I would also like to acknowledge the research team: Dr Isabel Metz, Ms Elizabeth Najdovski, Dr Susu Nousala and Mr Ordan Andreevski for their significant contributions.

A special acknowledgment to the case study co-authors: Isabel Metz, Danny Samson, B. Sebastian Reiche, Ordan Andreevski, Christopher Barnes, Susu Nousala, Suthida Jamsai, and Amy Lai.

Furthermore, I acknowledge the in-kind contribution provided by the respondents from each of the case study companies. Without their contribution, this book would not have been possible.

Finally, a special acknowledgment to Ms Yvonne Tan from World Scientific Publishing/Imperial College Press, whose advice and support made this book possible.

About the Author

Milé Terziovski is an Associate Professor of Innovation and Entrepreneurship in the Department of Management and Marketing at The University of Melbourne, where he teaches at undergraduate, postgraduate and executive levels. He has won the Dean's Award of Teaching Excellence for three consecutive years. He is also the Executive Director of the Centre for Global Innovation and Entrepreneurship at The University of Melbourne, where he has won three research grants from the Australian Research Council and two grants from the Department of Education Science and Training under the International Science Linkages Program to collaborate with European partners funded under the 5th and 6th EU Framework Programs.

After completing his PhD at The Melbourne Business School in 1996, he worked as a tenured Senior Lecturer in the Department of Management at Monash University. Prior to his academic career, Professor Terziovski worked for Rio Tinto for 16 years as an engineer in maintenance, production and project management. He has published more than 40 journal articles, 2 books and 2 in-press, 15 book chapters and more than 50 conference papers. He has published 3 articles in the prestigious *Journal of Operations Management*, winning the Best Paper award in 1999. His paper, based on his PhD thesis, has been in the top 10 most requested papers for the past five years. He has also been Guest Editor of the *International Journal of Manufacturing Technology Management* and has presented at the Academy of Management on several occasions, receiving a nomination for Best International Paper in 2000.

Professor Terziovski recently chaired the Innovation and Entrepreneurship Track at the European Academy of Management Conference (EURAM 2006) in Norway. He serves on two journal

Editorial Review Boards in the USA and two in the UK. He also served as an Advisor to the late President of the Republic of Macedonia, Mr Boris Trajkovski, on Innovation and Entrepreneurship, and is currently the Project Leader for the Global Manufacturing Research Group (GMRG) Project in Macedonia.

Contents

Preface... vii

Acknowledgments.. ix

About the Author.. xi

Chapter 1 Introduction.. 1

Milé Terziovski

1.1   Definitions of Innovation... 1
1.2   Evolution of Innovation.. 3
1.3   Innovation Capability.. 4
1.4   Enablers of Innovation Capability 6
1.5   Collaboration and Knowledge Transfer 10
1.6   Development of the Case Studies................................. 12
1.7   Overview of Book Chapters 14

**Chapter 2 Development of an Integrated Innovation
            Capability Model**... 19

Isabel Metz, Milé Terziovski and Danny Samson

2.1   Introduction 19
2.2   Objectives 20
2.3   Literature Review... 21
2.4   External Factors That Influence Innovation 22
2.5   Internal Factors That Influence Innovation.................. 25
2.6   Integrating e-Commerce, Sustainable
       Development, NPD and Innovation Capability 37
2.7   Sustainable Development (SD).................................... 40
2.8   Future Research Agenda 43
2.9   Synthesis of the Discussion....................................... 46
2.10   Conclusion 48

**Chapter 3 Strategic Shift from Product Orientation
to Innovative Solutions Capability in
the German Biotechnology Industry:**

Sartorius AG 51

Milé Terziovski and B. Sebastian Reiche

3.1   Introduction 51
3.2   Company Background 52
3.3   Business Strategy 54
3.4   Mission Statement 54
3.5   Core Competencies 55
3.6   Innovation Strategy 56
3.7   Resource Availability 57
3.8   Collaboration with External Partners and
       Absorptive Capacity 58
3.9   Complementary Assets 59
3.10   Innovation Capability 60
3.11   Knowledge Management 61
3.12   Sustainable Development 62
3.13   e-Commerce 64
3.14   New Product Development 65
3.15   Organizational Performance 66
3.16   Human Resource Management 68
3.17   Customer Orientation 69
3.18   Stakeholder Management 69
3.19   Conclusion 70

**Chapter 4 Managing Strategic Change Through
Mainstream and Newstream Innovation**

at Eurocopter, France 73

Milé Terziovski and B. Sebastian Reiche

4.1   Introduction 73
4.2   Case Study Interview 74
4.3   Company Background 75

4.4   Core Competencies.. 75
4.5   Innovation Approach 76
4.6   Resource Availability... 77
4.7   Mainstream and Newstream Capabilities 77
4.8   Stage 1: Innovation of the Blade Repair
        Process.. 79
4.9   The Role of Strategic Alliances................................... 81
4.10   Interaction of Mainstream and Newstream.................... 82
4.11   Human Resource Management 83
4.12   Conclusion 84
4.13   Implications for Managers.. 85

**Chapter 5 Leveraging Innovation Capabilities
        at Caterpillar Underground
        Mining (UGM) Pty Ltd** 87

Milé Terziovski and Ordan Andreevski

5.1   Introduction 87
5.2   Company Background 88
5.3   Perception and Definition of Innovation 90
5.4   Innovation Strategy 90
5.5   Core Competencies/Innovation Capability
        Development at UGM 91
5.6   Resource Availability... 94
5.7   The Role of Sustainable Development in Building
        Innovation Capability.. 95
5.8   The Role of e-Commerce in Building Innovation
        Capability 97
5.9   The Role of New Product Development (NPD)
        in Building Innovation Capability 99
5.10   Organizational Performance.. 100
5.11   Leadership and Culture.. 101
5.12   Lessons Learned and Opportunities for
        Improvement 103
5.13   Conclusion 103

**Chapter 6 Drivers of Innovation Capability at
        Sun Microsystems (SMS)**..................................... 105

Milé Terziovski and Christopher Barnes

6.1   Introduction 105
6.2   Company Background 106
6.3   Corporate Strategy.. 107
6.4   Mission Statement.. 108
6.5   Core Competencies.. 109
6.6   Innovation Strategy... 109
6.7   Resource Availability and Absorptive Capacity.............. 111
6.8   Innovation Capability.. 112
6.9   The Role of e-Commerce in Building Innovation
        Capability 114
6.10   The Role of New Product Development (NPD)
        in Building Innovation Capability 115
6.11   Organizational Performance 117
6.12   Opportunities for Improvement.................................... 117
6.13   Conclusion 118

**Chapter 7 Development and Exploitation of
        Innovation Capability at a Defence
        Project Engineering Company (DPEC)**.............. 121

Susu Nousala and Milé Terziovski

7.1   Introduction 121
7.2   Company and Industry Background............................... 121
7.3   Company Characteristics.. 122
7.4   Perception and Definition of Innovation 123
7.5   R&D Department... 123
7.6   Corporate Strategy... 124
7.7   Organizational Innovation Capability............................ 124
7.8   Manufacturing and Services 128
7.9   Sustainable Development (SD)....................................... 129
7.10   e-Commerce (e-Communication) 130

7.11 New Product Development (Through Project
      Management) 131
7.12 Self-Assessment and Continuous Improvement 132
7.13 Key Lessons Learnt 133
7.14 Organizational Performance 135
7.15 Opportunities for Improvement 136
7.16 Conclusion 138

Chapter 8 Drivers of Innovation Capability for Effective Sustainable Development: Best Practice at Vaisala 141

Milé Terziovski and B. Sebastian Reiche

8.1   Introduction 141
8.2   Company Background 142
8.3   Core Competencies 144
8.4   Mission Statement 145
8.5   Resource Availability 145
8.6   Innovation Strategy 145
8.7   Innovation Capability Model 147
8.8   Drivers of Innovation Capabilities at Vaisala 147
8.9   Integration of Innovation Capabilities 152
8.10 Supporting Capabilities 153
8.11 Conclusion 154

Chapter 9 Developing Innovation Capability Through Intellectual Property Strategy in the Australian Biotechnology Industry: Starpharma 157

Milé Terziovski and Amy Lai

9.1   Introduction 157
9.2   Innovation Capability and Commercialization
      Success 158
9.3   Biotechnology in Australia 158

9.4   Background to Starpharma............................ 160
9.5   Corporate Structure and Business Strategy................... 160
9.6   Workforce and Culture....................... 162
9.7   Core Activities, Products and Services........................ 163
9.8   Intellectual Property Strategy....................... 164
9.9   Alignment with Business Strategy............................ 164
9.10   Protection and Management........................ 165
9.11   Networks and Collaborations........................ 167
9.12   Resource Allocation 168
9.13   Innovation Capability........................ 169
9.14   Implementation of Protection and Management 170
9.15   Systems and Information Technology....................... 172
9.16   Conclusion 173
9.17   Implications for Managers.............................. 174

**Chapter 10 Development of Innovation Capability
              at Invincible Company in Thailand** 177

Suthida Jamsai, Susu Nousala and Milé Terziovski

10.1   Introduction 177
10.2   Company Background 179
10.3   Company Strategy 180
10.4   Current Level of Performance........................ 181
10.5   Invincible Customers 181
10.6   Product Innovation........................ 182
10.7   Product Innovation Process 185
10.8   Conclusion 188

**Chapter 11 Multiple Cross-Case Analysis:
              Conclusions and Implications** 191

Milé Terziovski

11.1   Introduction 191
11.2   Vision and Strategy......................... 192
11.3   Harnessing the Competence Base 195

11.4 Leveraging Information and Organizational
      Intelligence — Absorptive Capacity 200
11.5 Possessing a Market and Customer Orientation 204
11.6 Creativity and Idea and Knowledge Management 207
11.7 Organizational Structures and Systems 210
11.8 Culture and Climate 213
11.9 Management of Technology and Its Use 216
11.10 New Product Development (NPD) 219
11.11 Sustainability 222
11.12 e-Commerce 225
11.13 Characteristics of an Innovative Organization 227
11.14 Innovation-Driven Organizations: The Role of
      NPD, SDO and e-Commerce 232
11.15 Conclusion 233
11.16 Implications for Managers 234

References 237

Index 249

Chapter 1

Introduction

Milé Terziovski

1.1 Definitions of Innovation

The study of innovation appears in different literatures such as sociology, education, management, etc. In the management literature, two schools of innovation research have been identified by Subramanian and Nilakanta (1996). The first domain is that developed by marketing researchers who are interested in understanding the causes of innovative behavior of consumers, where the consumer is used as the unit of analysis. The second domain of innovation management research is that developed by researchers in areas of organizational theory and strategic management, where the organization is used as a unit of analysis. We have chosen the second domain for the purpose of this study. There are many definitions of innovation in the literature. Damanpour (1991) defines innovation from an organizational perspective:

> the adoption of an idea or behavior, whether a system, policy, program, device, process, product or service, that is new to the adopting organization.

Porter and Stern (1999, p. 12) define innovation from a customer perspective as

> ... the transformation of knowledge into new products, processes, and services — involves more than just science and technology. It involves discerning and meeting the needs of the customers.

Others view innovation as an enabling device for producing new products and processes on a continuous basis (Dougherty & Hardy,

1

1996). Innovation is about using knowledge to offer a new product or service to customers via lower costs or improved attributes (Afuah, 1998). Such improvements can emerge from innovations that are either of a product or process nature. Adding to the dilemma of defining innovation, there are also questions of whether an innovation is of a radical or incremental nature.

Incremental innovations occur continuously in the organization and lead to minor improvements in products or processes. Jha *et al.* (1996) define continuous improvement (CI) as a collection of activities that constitute a process intended to achieve performance improvement.

In manufacturing, these activities primarily involve simplification of production processes, chiefly through the elimination of waste. In service industries and the public sector, the focus is on simplification and improved customer service through greater empowerment of individual employees and correspondingly less bureaucracy (Samson & Terziovski, 1999; Schroeder *et al.*, 2002). On the other hand, radical innovations are more long-term and strategic in focus, and aim to change key capabilities of the firm, thus creating a new operating paradigm. According to Harrington (1995), "all organizations need both continuous and breakthrough improvement."

However, according to Harrington, continuous improvement is the major driving force behind any improvement effort. Breakthrough improvement serves to "jump-start" a few of the critical processes. Several researchers are concerned that past research has focused on technological and technical product innovation to the neglect of process and organizational innovation. For example, Harvard Business School researchers Kim and Mauborgne (1999) integrate customer value with technology innovation under the term "value innovation." Innovation often is least effective when there is application without considering who will value the development, either as internal or external customers. Technology innovation on its own does not address buyer value, thus a new technology might not be accepted in the market as having value for the customer. Technology innovation tends to focus on specific solutions, whereas value innovation focuses on redefining and solving the problem which leads to a customized solution.

Many of the definitions of innovation discussed above are quite broad. A global definition of innovation does not exist; rather, different definitions of innovation are appropriate under different circumstances. A narrow definition of innovation may be a useful tool in researching the activities that lead to greater organizational performance. The following definition of innovation has been articulated by the Author:

> Innovation is the application of resources to create value for the customer and the enterprise by developing, improving and commercializing new and existing products, processes and services.

1.2 Evolution of Innovation

Since the 1930s, our view of what constitutes "innovation" has changed. Rothwell (1994) explains the evolution of innovation along five generations of behavior:

First generation innovation (1G) — technology push. This era of innovation was the foundation for the Industrial Revolution. Innovation came with new, technologically advanced products and means of production. Such products were pushed onto the market.

Second generation innovation (2G) — need pull. Innovation during this era shifted to a market/customer focus, a focus where the customer determined needs and production technology responded. Marketing took a pivotal role in generating new ideas.

Third generation innovation (3G) — coupling model. This era of innovation involved a coupling of the push and pull models. The market might need new ideas, but production technology refined them. Alternatively, R&D developed new ideas that marketing refined with market feedback. R&D and marketing were linked.

Fourth generation innovation (4G) — integrated model. An integrated model of innovation saw a tight coupling of marketing and R&D activity, together with strong supplier linkages and close coupling with leading customers.

Fifth generation innovation (5G) — systems integration and net-working model (SIN). This model of innovation builds on the inte-grated model by including strategic partnerships with suppliers and customers, using expert systems, and having collaborative marketing and research arrangements. There is an emphasis on flexibility and speed of development with a focus on quality and other non-price factors.

According to Rothwell (1994, p. 22), fifth generation (5G) innovation has both strategic and enabling characteristics. Strategic elements include time-based strategies (faster, more efficient product development); a development focus on quality and other non-price factors; an emphasis on corporate flexibility and responsiveness; a customer focus at the forefront of any strategy; a strategic integration with primary suppliers; electronic data processing strategies; and a policy of quality control. Enabling factors include a greater level of overall organization and systems integration; a flatter, more flexible organizational structure for rapid and effective decision-making; fully developed internal databases; and an effective external data link.

Rothwell (1994) claims that there are examples of Japanese firms operating on fourth generation innovation, and U.S. firms operating on third generation innovation, but the presence of fifth generation innovation is still emerging. The eight case studies discussed in this book fall between third and fourth generation innovation companies.

1.3 Innovation Capability

As innovation evolved from 1G to 5G, new enabling factors and drivers of innovation have also evolved: e-Commerce, sustainable development and a focus on accelerating new product development. These are taking leading roles in helping to transform knowledge into new products, processes and services. As organizations have down-sized and worked on cost reduction for many decades now, and sim-ilarly improved their quality and service, they have generally achieved efficiency and process stability outcomes. The next battlefield that will drive international competitiveness and business outcomes of firms is innovation. Therefore, the burning question addressed in this book is:

> What constitutes innovation capability in organizations, and how can it be developed and exploited?

Despite the volumes of research on innovation in organizations, there are no clear, agreed guidelines for creating innovation-driven organizations. Numerous studies have attempted to isolate the important variables facilitating innovation outcomes (Damanpour, 1991). However, there is still much we do not know about how firms can innovate faster and better. We do know that effective innovation requires the construction of an overarching framework of factors conducive to creativity (Kanter, 1989). The absence of such frameworks may lead to a conservative and ineffective innovation culture. Lawson and Samson (2001) define innovation capability as:

> ... the ability to continuously transform knowledge and ideas into new products, processes and systems for the benefit of the firm and its stakeholders.

Innovation capability provides the potential for effective innovation. However, it is not a simple or single-factored concept, as it involves many aspects of management, leadership and technical aspects as well as strategic resource allocation, market knowledge, organizational incentives, etc. Lawson and Samson (2001) identified several dimensions of innovation capability which are listed below:

- Vision and strategy;
- Harnessing the competence base;
- Leveraging information and organizational intelligence;
- Possessing a market and customer orientation;
- Creativity and idea management;
- Organizational structures and systems;
- Culture and climate;
- Management of technology.

Our aim is to explore each of the organizational innovation capability dimensions above to identify which of these individually or in combination are perceived by best practice innovative organizations as critical to effective innovation. Therefore, an effective organizational design encompassing systems, structures, rewards and strategy will increase the probability of generating new ideas and support their journey toward commercialization.

1.4 Enablers of Innovation Capability

Current and future challenges and opportunities facing business and government organizations are in the fields of sustainable development; e-Commerce; and new product development. Recent work by Porter and Stern (1999) has shown the three identified domains above to be of critically important interest to today's governments and many organizations.

The power of the innovation capability construct is that it is generalizable to all these domains, as it relates to the organizational potential to convert new ideas into commercial and community value. Developing innovation capability in these three domains provides valuable insights as to how the innovation capability construct can be further developed, and will also provide valuable practitioner guidelines. The elucidation and validation of the innovation capability construct in these three fields would add significantly to the bodies of knowledge in each of the three fields, and in the central innovation management literature.

Sustainable Development

Sustainability has clearly begun to assert itself as a driver for innovation. Organizations need to better understand how the emergence of environmentalism and sustainable development impacts on firm's opportunities and capability to innovate.

Ottman and Reilly (1998) suggest that firms have responded and profited with the emergence of environmentalism as a core societal value. "Green" marketing is increasingly being seen as an opportunity for innovation. Firms require knowledge on how to create new products, how to identify and capitalize on opportunities to innovate, and how to communicate effectively.

Polonsky (2001) argues that going green provides a firm with strategic advantages including lower costs, differentiation and revitalization. Gertakis (2001) has illustrated how the new product design process can integrate environmental factors within a commercial context. Many of these environmental technologies are more widespread in cleaner production and pollution prevention; however, their incorporation according to Gertakis into products is not as extensive. Gertakis cites a number of exceptions to this norm.

For example, Kambrook developed a kettle that has improved energy efficiency and is designed to facilitate disassembly and recycling. Blackmores has redesigned its packaging to reduce material consumption. Dishlex has designed a dishwasher that uses less water, has improved energy efficiency, has reduced material consumption, is "light-weighted" and is designed to facilitate disassembly and recycling. This leads to the question:

> How does innovation capability manifest in the sustainable development innovations?

e-Commerce

The arrival of e-Commerce has driven firms to re-evaluate their entire way of doing business, and in many cases, create entirely new forms of competition. e-Commerce acts as both a driver and enabler of innovation within organizations. As a driver of innovation, e-Commerce has underpinned stronger, more rapid and flexible competition forcing firms to restructure competitive boundaries and re-evaluate existing practices, products and services.

As an enabler of innovation, e-Commerce provides immense scope for organizations to discard old processes, diffuse local innovations globally, remove constraints to innovation and create entirely new innovative practices and models. These companies use e-Commerce technologies to profit from their intellectual capital and to evolve at the same or greater pace than the market.

However, many organizations, particularly small to medium enterprises (SMEs) are struggling with the initial stages of e-Commerce policy. It is the lack of an integrated strategy and resources that restricts many organizations from realizing the full potential of this new form of organization. The "digital divide" between large, highly resourced companies and SMEs is a significant impediment to SME performance improvement.

Callahan and Pasternack (1999) in Wheelen and Hunger (2004) report the results of a study they conducted on a sample of senior executives in the USA to identify the impact of the internet on the future organizational model and concluded the following:

(1) The internet is forcing companies to transform themselves. The concept of electronically networking customers, suppliers and partners is now a reality.
(2) New channels are changing market access by working directly with the customers.
(3) The balance of power is shifting to the consumer. Customers are much more demanding, having unlimited access to information on the internet.
(4) Competition is changing. New technology-driven firms plus older traditional competitors are exploiting the internet to become more innovative and efficient.
(5) The pace of business is increasing drastically. Planning horizons, information needs, and customer supplier expectations are reflecting the immediacy of the internet.
(6) The separation between suppliers, manufacturers and customers is becoming blurred with the development and expansion of extranets, in which cooperating firms have access to each other's internal operating plans and processes.
(7) Knowledge is becoming a key asset and a source of competitive advantage. For example, physical assets accounted for 62.8 percent of the total market value of US manufacturing firms in 1980 but only 37.9 percent in 1991. The remainder of the market value is composed of intangible assets, primarily intellectual capital (Kanter, 1999; Allee, 2000).

The above discussion on e-Commerce leads to the question:

How does the innovation capability manifest in the new economy (e-Business) firms?

New Product Development

Research on NPD indicates that many of the factors critical to innovation in general are also linked to successful NPD. In particular, factors linked to accelerating the NPD process, such as cross-functional teams and external cooperative relationships, may reflect capabilities specific to innovative organizations. Mabert *et al.* (1992), based on a comparison of six NPD projects, concluded that a knowledgable leader with sufficient time to devote to the management of the project, shorten the development time of new products.

Most empirical work on NPD has focused on the relationships between various success factors, including new product strategies, and performance measures and risk. As a result, we know for example that firms which emphasized market innovativeness in their product introductions enjoyed higher returns than those which did not (Firth & Narayanan, 1996, p. 334). Past research also indicates that critical factors of NPD are, for instance, a clear, well-communicated new product strategy; strategic focus and synergy; an entrepreneurial climate for product innovation; adequate resources for new products; senior management commitment to and accountability for new product development; and the existence of high-quality, cross-functional development teams in the organization (Powell *et al.*, 1996).

In addition, motivation triggered by competitive pressures, aspects of teamwork such as full-time participation and cross-functionality of teams, outside influences such as vendor participation in the project, and systematic project control, may accelerate the NPD process (Mabert *et al.*, 1992; Sohal *et al.*, 2003).

In order to accelerate the innovation process one needs to understand what factors are critical to the successful execution of the cycle. Omta *et al.* (1997) have demonstrated that if organizations are to be "innovative", management face the challenge of creating conditions conducive to meeting the goals of scientific performance as well as the scientists' need for satisfaction and motivation (socio-technical performance). According to Omta (1995),

> The best performing companies in [pharmaceutical] development are able to shorten the development phase by more than a year, by use of parallel development and close monitoring of the developmental process. In the more-than-average performers fine-tuning is more precise, and the lateral and cross-functional communication more intense leading to a concurrent process.

Managers need to acquire new skills and management systems to assist them in managing the "knowledge boundaries" of their firms, increasing innovativeness and deriving benefit from strategic alliances and partnerships. Most biotech firms are networked through partnerships, alliances, formal and informal collaborations, and agreements (Koput *et al.*, 1996).

> How does innovation capability manifest in the new product development processes in the biotechnology industry?

1.5 Collaboration and Knowledge Transfer

The need to be simultaneously efficient, flexible and adaptive has accelerated the evolution of the network form of organization. Networks composed of multiple specialist companies as their key building blocks have been called "modular corporations" (Miles & Snow, 1992). Multi-firm networks that change their shape often and quickly have been called "virtual corporations." Many network organizations have also been formed in mature industries because older established companies came to realize that they were too large and cumbersome to respond effectively to competitive demands of today's environment. For example, General Electric was among the first of American companies to restructure in the early 1980s.

The overall objective of the restructuring is to reduce the centralized coordination requirements and create the flexibility necessary to get close to customers and the speed required to meet their demands in a timely fashion. The restructured companies establish more and smaller business units, form cross-functional teams that are responsible for key processes, and design reward schemes to encourage entrepreneurial behavior on the part of their managers and employees.

Previous research by Soderquist (1996) has identified the "development of networks and partnerships" as a critical success factor (CSF) for improving competitiveness in SMEs. An important method to keep the information and knowledge flowing into SMEs is by locating within an SME cluster. Being able to leverage limited resources by establishing collaborative relationships with similar organizations has proven to be successful in many parts of the world (Baptista & Swann, 1998). This strategy can also be coupled with strategic partnering to network in needed skills and expertise lacking internally.

One of the most frequently cited examples of networking among SMEs involves the "industrial districts" in Italy (Miles & Snow, 1992). Thousands of small firms specializing in various trades both compete and collaborate with each other. For example, one network of 15 small engineering firms formed partnerships in order to develop collective clout in the marketplace. However, each firm was able to remain as a separate legal entity with its own workforce, facilities, accounting systems, etc.

An American variant to the Italian industrial districts is seen in California's Silicon Valley. The primary motive of these firms in forming networks is to obtain the advantage of "bigness" while remaining small. By remaining small, each network firm in businesses such as biotechnology, semiconductors and a host of others can be highly responsive. It has fewer bureaucratic procedures that must be overcome in order to respond to requests from customers or network partners. Furthermore, each small network firm is a specialist in a particular technology. Usually it is at the leading edge of its area of expertise and is therefore a prime candidate when other firms need a certain type of expertise.

On the other hand, large firms focus on those core competencies in which they can compete on a world-class basis, and they outsource remaining activities to upstream or downstream partners, usually

SMEs. Alternatively, organizations can connect with networked incubators and invest in new ventures that can drive the entrepreneurial sprit into a mature organization (Hansen *et al.*, 2000). With over 350 incubators announced worldwide over the last several years, this emerging organizational structure provides a powerful opportunity to connect with fledgling companies to bring new ideas to market faster. While a mature organization investing in a start-up can provide valuable lessons to young entrepreneurs and a progression path for their new ideas, the challenge lies in allowing the incubator free range and not stifling their creativity with bureaucratic processes.

Terziovski (2003) conducted a cross-sectional study of Australian SMEs. Quantitative data was gathered from a stratified random sample of SME site managers in the Australian manufacturing industry. A total of 550 questionnaires were sent to manufacturing managers from which a response rate of 20 percent was achieved. Networking practice models were developed in order to test the strength of the relationship between key components of networking practice and several dimensions of Business Excellence such as success rate of new products, reduction in waste, quality management, etc.

The data was analyzed using techniques available on the SPSS for Windows software package. Terziovski (2003) concluded that groupings of network practices are required to explain Business Excellence. This means that a single networking practice is not sufficient to explain Business Excellence significantly. The networking practice that has the most significant explanatory power was found to be the establishment of formal support systems such as communication linkages within networks.

The main implication of the research results for SME managers is that a typical manufacturing SME is more likely to improve Business Excellence with a combination of networking practices than without these practices.

1.6 Development of the Case Studies

Qualitative data was gathered using a case study protocol, which was designed specifically for the case study research. An analysis of events during a three-year period are documented in terms of what happened, why it happened, how it happened, who was involved and the main lessons learnt. The multiple cross-case analysis approach was selected to seek and explain "best practice" implementation of innovation. Multiple case study designs offer the advantage of more information over single case study designs.

However, they also demand more resources and time. According to Yin (1989), the multiple case study design allows "replication" logic. This is the logic of treating a series of cases as a series of experiments, where each case study serves to confirm or refute the conclusions drawn from the previous ones (McCutcheon & Meredith, 1993). Written analysis of multiple case studies may take three well-known forms as outlined by Balan (1994): narrative of each case study to describe and analyze the information; narrative form with multiple cross-case analysis; and where the entire discussion consists of the cross-case analysis. The narrative form with multiple cross-case analyses is adopted in this book. The qualitative research addresses an issue in management that has been discussed by various disciplines but very little integrative research has been conducted in a comprehensive manner. The case studies collectively provide significant new knowledge to the existing research knowledge base and to practitioners in the organization's "mainstream" and the "newstream" as proposed by Kanter (1989).

The qualitative research findings provide an in-depth understanding of key practices that play a significant role at the various stages of the innovation process. Managers receive the rhetoric via government policy and management theory that innovation is key to continued success. Often the advice focuses on simply telling organizations to provide a larger research budget and to protect their innovations through patents, etc. Research results do not provide any consistency across industry or firm size into the management of both product and process innovation (Wolfe, 1994; Afuah, 1998). Based on the above discussion, the research question motivating this book is:

What constitutes innovation capability in organizations, and how can it be developed and exploited? What are the key drivers of innovative organizations?

1.7 Overview of Book Chapters

Chapter 2 — Development of an Integrated Innovation Capability Model

The unique contribution of this chapter is the development of an integrated innovation capability model (IICM) grounded in the literature. Chapter 2 shows how the innovation capability construct is generalizable to all three domains of new product development, sustainable development orientation and e-Commerce, as it relates to the organizational potential to convert new ideas into commercial and community value. Based on an extensive literature review, the authors conclude that innovation capability can be enhanced through expertise gained in the three domains of NPD, SDO and e-Commerce. Also, common elements of innovation capability bring value to the three domains. Finally, synergistic innovation capability effects are obtained from the integration of the three domains.

Chapter 3 — Strategic Shift from Product Orientation to Innovative Solutions Capability in the German Biotechnology Industry: Sartorius AG

This chapter shows how Sartorius achieved strategic shift from product orientation to innovative solutions capability. Sartorius AG is a leading company in the field of biotechnology and mechatronics. Its strategic shift in its business model from a product-oriented firm to a total solution provider has been a major success factor for maintaining its market position and continuously satisfying its customers through a systematic anticipation of their potential needs based on the company's technology portfolio. The case analyzes key components of the organization's innovation capabilities and examines the company's innovation strategy. Sartorius' main innovation capabilities lie in the systematic integration of customers into the product development process. The authors conclude that innovation needs to be strategic in nature in order to sustain firm-specific competitive advantage.

Chapter 4 — Managing Strategic Change Through Mainstream and Newstream Innovation at Eurocopter, France

This chapter shows how Eurocopter adopts an integrated innovation perspective by re-configuring its core business processes (mainstream) and systematically involving its customers in the product development in pursuit of increased customer value (newstream). The qualitative analysis indicates how the company establishes a strategic network of external partners to tap into additional sources of innovation and thereby enhance its innovation capabilities. Customer involvement has become part of a more far-reaching change in customer philosophy at Eurocopter. The case study findings have important implications for managers. The need for managers to reach beyond their immediate organizational boundaries and find additional sets of resources that can support both product and process innovation is one of the main implications.

Chapter 5 — Leveraging Innovation Capabilities at Caterpillar Underground Mining (UGM) Pty Ltd

This chapter looks at the Australian operation of Caterpillar Underground Mining Pty Ltd (referred to in the case study as UGM) in Burnie, Tasmania, with particular emphasis on how UGM has nurtured its innovation capabilities to achieve a dominant market share and brand positioning. The case examines the relationship between innovation enablers, innovation capability and innovation performance. The analysis demonstrates that sustainable innovation needs to be embedded into the corporate culture, business models and practices, and the process is ongoing and systemic.

Chapter 6 — Drivers of Innovation Capability at Sun Microsystems (SMS)

This chapter shows how SMS develops competitive advantage through an integrated focus on vision, mission, customer focus, innovation capability, cooperation, commitment to quality service and making computer power more affordable. SMS has taken advantage of the innovation enablers like new product development, e-Commerce and sustainable development to continuously improve its innovation performance and sustain it innovation capability which is the principal source of its competitive advantage.

Chapter 7 — Development and Exploitation of Innovation Capability at a Defence Project Engineering Company (DPEC)

This chapter develops an understanding of how innovation capability is developed at DPEC and how knowledge is transferred in the product innovation process. The qualitative analysis reveals that the process of developing and applying innovation knowledge occurred in two different phases. The first phase built up new knowledge through various innovation practices, such as the design of original products by small divisional teams (SDTs). The second phase captured existing innovation knowledge through reflection, learning and understanding of key steps throughout the process of innovation. Knowledge transfer was seen as critical in reducing the NPD lead time, so as to meet the needs of the customer and the organization in terms of time, cost and quality.

Chapter 8 — Drivers of Innovation Capability for Effective Sustainable Development: Best Practice at Vaisala

This chapter reports the results of a qualitative case study at the Melbourne subsidiary of Vaisala, examining the role of drivers of innovation capabilities for effective innovation output. We draw upon an integrative framework of innovation capabilities that proposes both a single and interactive effect of sustainable development, e-Commerce and new product development on a firm's innovation capabilities. The analysis of the case study data reveals that Vaisala's innovation capabilities rest to a considerable extent on the integration of the three driving factors specified in the conceptual model. The firm's streamlined structure, highly-skilled work force and high knowledge and research intensity serve as supporting capabilities that help to interlink these enabling factors. Implications for innovation practice and research are discussed.

Chapter 9 — Developing Innovation Capability Through Intellectual Property Strategy in the Australian Biotechnology Industry: Starpharma

This chapter explores Starpharma's strategic view of innovation and how intellectual property contributes to innovation capability to achieve commercial success. Starpharma has a broad view of intellectual property, which encompasses codified and tacit knowledge as well as people and relationships which all simultaneously contribute to the development of innovation capability. The company has the view that without human and fiscal resources to commercialize intellectual property, there would be little value to the organization. The company believes that it is important for its corporate culture to be diffused with values based on intellectual property, as long as these are part of building innovation capability which is the critical source to Starpharma's sustainable competitive advantage.

Chapter 10 — Development of Innovation Capability at Invincible Company in Thailand

This chapter explores the relationship between new product development and e-Business at Invincible Company. The company developed innovation capability through strategies that involved understanding the emergence of innovative ideas and the evolution of specific practices in managing innovation capability. Key successes to innovation included approaches such as the "we do better" strategy, which replaced traditional processes with more supportive ones. Other important aspects undertaken allowed the manager to combine two styles of management, flexible (informal) and formal, to form a new strategy. One of the key findings in this case study was the way in which NPD and e-Business activities were related to the innovation process and how they were applied.

Chapter 11 — Multiple Cross-Case Analysis: Conclusions and Implications

This chapter provides the multiple cross-case analysis of the eight case studies. A balance between "hard" and "soft" innovation practices is necessary for innovation to be successful and sustainable both in the "mainstream" and in the "newstream." The key drivers of innovative organizations were found to be: committed leaders, a highly developed innovation strategy, a "first-to-market" philosophy of new products and services; supported by effective "top-down" and "bottom-up" communication processes. New Product Development strategy under-pinned by cross-functional teams, e-Commerce and Sustainable Development Orientation were identified as effective enablers of highly innovative organizations. The set of sustainable development practices and the firm's approach to SD are likely to depend on the industry the organization is in. In turn, internal factors such as man-agement style, technology and the firm's financial position are likely to influence the firm's approach to sustainable development.

Review Questions

(1) Why is innovation such a difficult concept to define? Discuss the importance of innovation in driving international competitiveness.
(2) Articulate the factors that have driven the evolution of innova-tion from 1st generation to 5th generation innovation.

Chapter 2

Development of an Integrated Innovation Capability Model

Isabel Metz, Milé Terziovski and Danny Samson

2.1 Introduction

The ability to continuously innovate is of critical importance to the long-term success of the organization. Yet, it is not clearly understood what constitutes innovation capability in an organization, and how it can be developed and exploited. Teece, Pisano and Shuen (1997) define "dynamic capabilities" as:

> ... the firm's ability to integrate, build, and reconfigure internal and external competences to address rapidly changing environments (p. 516).

Lawson and Samson (2001) define innovation capability as:

> the ability to continuously transform knowledge and ideas into new products, processes and systems for the benefit of the firm and its stakeholders (p. 8).

Thus, innovation capability is interpreted as a combination of factors, internal and external to the organization, which are linked to the organization's ability to continuously innovate. Innovation capability is therefore perceived as a complex concept because it can be influenced by factors internal and external to the organization, such as leadership capabilities (e.g., Ahmed, 1998; Angle, 1989; Damanpour, 1991; Souder, Song & Kawamura, 1998) and the level of industry innovativeness (e.g., Audretsch, 1995), respectively. As pressures for

19

innovation have increased, new enabling factors and drivers have also evolved: e-Commerce, sustainable development and a focus on accelerating new product development are taking leading roles in helping to transform knowledge into new products, processes and services. These three elements can have a role in the innovation capability of organizations.

For example, sustainable development concerns can propel organizations to improve their processes and products (e.g., Dunphy, Griffiths & Benn, 2003). New product development processes can contribute to an organization's innovation management capability. e-Commerce can drive communication and networking effectiveness forward, both inside the firm and across its boundaries. Both e-Commerce and sustainable development are relatively recent concerns for organizations. Both, however, have the potential to provide organizations with huge opportunities or costs (Dunphy *et al.*, 2003; Samson, 2003).

The evidence that exists indicates that most organizations worldwide have yet to effectively and innovatively combine e-Commerce and sustainable development practices with traditional business models and approaches (Dunphy *et al.*, 2003; Samson, 2003). Therefore, there is a need to understand the roles of both e-Commerce and sustainable development, particularly as contributing factors to innovation capability. In addition, not only has the process of NPD been closely associated with innovation (e.g., Cooper, 1985; Damanpour, 1991; Dougherty & Hardy, 1996), but indeed the process of accelerated NPD is considered increasingly critical to the competitiveness of organizations (e.g., Mabert *et al.*, 1992; Pawar & Sharifi, 1997). Therefore, there is a need to understand the role of NPD more generally within innovation capability *vis-à-vis* those of e-Commerce and sustainable development. As a result, the first aim of this article is to address the research question: *What are the characteristics of innovation capability from the existing literature?*

2.2 Objectives

The question articulated above is addressed by a review of the existing literature on predictors of success in innovation, e-Commerce, sustainable development and NPD and conducting a meta-analysis of the literature in order to identify the factors that are potentially common to all four areas of study. Based on the analysis, an innovation capability model is developed and future research agenda is articulated. This meta-analysis is necessary because numerous studies have attempted to isolate the important variables facilitating innovation outcomes (e.g., Ahmed, 1998; Angle, 1989; Damanpour, 1991), but each has focused on part of the whole picture.

In particular, to the authors' knowledge, there are no studies that integrate the more traditional notions of innovation with the roles of e-Commerce, sustainable development, and NPD to explain innovation capability. Therefore, the second aim is to set the research agenda for the next several years in innovation management, so that new theories and practical insights for practitioners in the field of innovation capability can emerge. We conclude by presenting an integrative view of what can constitute innovation management capability, by combining the predictors of the traditional notion of innovation with organizational capabilities in e-Commerce, sustainable development and NPD.

2.3 Literature Review

A common "input-based" notion of innovation is the amount invested in research and development (R&D) (e.g., OECD, 2000), but other notions exist. For example, Damanpour (1991) defines innovation as the "adoption of an internally generated or purchased device, system, policy, program, process, product, or service that is new to the organization" (p. 556). In addition, Porter and Stern (1999) explain that innovation "involves discerning and meeting the needs of the customers" (p. 12). Damanpour's definition encompasses the concept of open innovation.

According to the model of open innovation, "firms commercialize external (as well as internal) ideas by deploying outside (as well as in-house) path ways to the market" (Chesbrough, 2003, pp. 36–37). In turn, Porter and Stern's (1999) definition encompasses the customer, rather than only external factors such as science and technology. In line with the aims stated earlier (e.g., Teece *et al.*, 1997), this review draws on the literature that examines the internal and external links of the firm, including links between innovation and government regulation, industry participation, organization characteristics, and the management of human resources. In addition, this review also draws on the literature on the firm's competences in e-Commerce, environmental sustainability, and accelerated NPD to understand their relationship to innovation and performance.

2.4 External Factors That Influence Innovation

In this section we review the influence of government regulation, industry type and characteristics, customers and competitors, and partnerships on an organization's ability to innovate.

Government Regulation

Government regulation can provide firms with opportunities or constraints, depending on the context and how they approach it. Although government regulation is not the only factor that can positively impact on the process of innovation, it is nevertheless an important one (Delaplace & Kabouya, 2001; Dunphy *et al.*, 2003; Porter & Stern, 1999). Regulatory factors that can influence innovation range from tax to patent and copyright laws (Porter & Stern, 1999). However, the interactions between environmental and e-Commerce regulation, and firm and industry innovation in these two areas are examined below as key relationships in this regard.

Environmental Regulation

An enterprise's sustainable development orientation (SDO) reflects the influence of and ability to influence environmental, political and social factors. An organization's SDO can be seen as "the degree to which the organization culture and its set of SD practices are efficient and effective both in meeting economic, environmental and social needs and in supporting the strategic direction of the business, hence providing greater opportunity for long-term superior business success" (Goldsmith & Samson, 2002, p. 15). For example, stringent environmental regulations in the developed countries have been critical in getting organizations to focus on sustainability (e.g., Delaplace & Kabouya, 2001; Hart, 1997).

For example, government regulation on car emissions forced the auto industry to focus on pollution control (Hart, 1997). Therefore, it can be concluded that government regulation on the environment can have an impact on the SDO of the firms in that industry. In fact, scholars are of the opinion that more needs to be done to encourage whole industries to have a vision for sustainable development that will ensure the innovative use of new technologies and full product lifecycle design to minimize waste and pollution (e.g., Hart, 1997).

e-Commerce Regulation

Research indicates that e-Commerce regulation is necessary for e-Commerce to flourish. For example, Debreceny, Putterill, Tung and Gilbert (2002) concluded that lack of user trust in the security of the transactions undertaken over the internet is one of the main inhibitors of e-Commerce. In addition, the participants in Debreceny *et al.*'s (2002) study felt that the intergovernmental framework in place for the transfer across borders of goods and services was inadequate, and standardization of terms and conditions in the Asia-Pacific region was required. Yet, Debreceny *et al.* (2002) concluded that aspects of public policy, such as security and infrastructure, are less important as e-Commerce inhibitors than past researchers had indicated. Debreceny *et al.*'s (2002) conclusion needs to be interpreted with care, because they did not explain all the e-Commerce inhibitors found in their study, nor did they examine the relative importance of those inhibitors to the adoption of e-Commerce.

Industry

The industry that an organization operates in can influence its ability to successfully innovate. For example, Holak, Parry and Song (1991) found that the relationship between R&D and a firm's performance, for firms in the industrial-goods market, was influenced by industry characteristics such as the recentness of the technology change, customers' reliance on professional advisors, and the purchase price of the product. In addition, Audretsch (1995) found that in innovative manufacturing industries, the rate of survival of new entrants or small firms was lower than in non-innovative industries. Yet, as firms survived the first few years in innovative industries, being in an innovative industry increased their likelihood of survival. Further, organizations in more innovative industries were likely to have greater innovation capabilities than firms in less innovative industries. Furthermore, Acs and Audretsch (1988) concluded that the industries in which small firms have an innovation-based advantage tend to be different from those of large firms. Factors that affect the innovative advantage of small versus large firms can be, for instance, how difficult it is to copy the knowledge of existing firms, and how quickly and radical the changes are in technology in a particular industry. The type of industry and the degree to which an industry is innovative is likely to influence the innovation capabilities, and ultimately the survival, of the firms in that industry.

Customers and Competitors

Customers and competitors in a particular industry can influence the innovation capability of organizations in that industry. For example, technologically sophisticated customers may demand more innovative products and services (Porter & Stern, 1999). In addition, geographically close competitors may be able to identify customer needs and source components more quickly than isolated ones (Porter & Stern, 1999). As a result, customers and competitors may influence the innovative capability of organizations.

Partnerships

Porter and Stern (1999, p. 17) found that "clusters" were strongly and positively related to national innovativeness. Clusters were defined as

> geographically proximate groups of interconnected companies, industries, and associated institutions in a particular field, linked by commonalities and complementarities.

It is possible that clusters are also linked to the innovation capability of organizations. In support, cooperative relationships with customers, suppliers, vendors (e.g., Dunphy *et al.*, 2003; von Hippel, Thomke & Sonnack, 1999; Mabert, Muth & Schmenner, 1992; Terziovski, 2003) and government authorities (Dunphy *et al.*, 2003; Sharfman, Meo & Ellington, 2000) appear to enable innovation.

In some instances, cooperative relationships with competitors, or co-opetition, result in better innovation performance than competitive strategies. This may be because not any one stakeholder has all the expertise and resources necessary to successfully innovate. Different stakeholders are also likely to play different roles in innovation, as illustrated by the analyses of highly successful and innovative companies such as 3M (von Hippel *et al.*, 1999).

2.5 Internal Factors That Influence Innovation

Generally speaking, past research indicates that a firm's characteristics influence some relationships commonly associated with innovation. For example, Holak *et al.* (1991) found that the positive relationship between R&D and a firm's performance is stronger for firms that provide high quality after-sales customer service than for firms that do not, and that this relationship is different for growth-stage and mature firms. Therefore, in this section we review the literature on the relationship between innovation and the organizational characteristics of size, strategy, structure, the management of technology, and market knowledge, amongst others.

Organizational Size

The relationship between organization size and innovation is complex, as indicated by Damanpour (1992). In a meta-analysis of 20 empirical studies, Damanpour (1992) concluded that size was positively related to innovation but some factors, such as the type of organization and the stage of adoption, moderated this relationship. For instance, size was "more positively related to innovation in manufacturing than in service, and in profit, rather than in non-profit organizations" (Damanpour, 1992, p. 385). In turn, Subramanian and Nilakanta (1996) found that organizational size was generally associated with technical and administrative innovativeness.

Furthermore, the OECD *Policy Brief* (2000) states, "large firms are more involved in technological alliances than small firms" (p. 6). Yet, "small start-up firms are more flexible and unencumbered than large established firms and are essential to the 'creative destruction' that occurs in periods of technological change" (p. 7). In addition, Theyel (2000) found a relationship between plant size (of firms in the US chemical industry) and the environmental practice of conducting waste audits. Furthermore, Konings and Roodhooft (2002) found that large firms are more likely than small firms to use and benefit (in terms of productivity) from e-Commerce.

Consequent to the above discussion, organization size does not appear to be linked in the same manner to all dimensions of innovation nor to all environmental management practices. The relationship between organization size and the organization's ability to innovate appears to be moderated by factors such as the measure of size, the scope of innovation, the type of organization, and the stage of adoption (Damanpour, 1992). However it is clear that large and small firms have different types of opportunities to innovate, and, hence it is possible to conclude that size does matter to innovation capability.

Strategy

A business strategy addresses the question of how the company or its business units can compete in its businesses and industries (Wheelen & Hunger, 2002). In general, research has shown that effective strategic management can assist firms to outperform their competitors. Strategic management is "that set of managerial decisions and actions that determines the long-term performance of a corporation" (Wheelen & Hunger, 2002, p. 2).

In particular, a firm's strategy regarding NPD (e.g., Cooper, 1985; Cooper & Kleinschmidt, 1996), new technology adoption or e-Commerce (e.g., Caldeira & Ward, 2003; Porter, 2001), sustainable development (e.g., Dunphy *et al.*, 2003; Hart, 1997), and the more general concepts of innovation (e.g., Ahmed, 1998; Terziovski, 2001, 2002) appears to be a critical factor in the firm's success. In addition, different strategies are likely to influence different types of innovation and performance.

For instance, Terziovski (2002) found that a continuous improvement strategy most explained productivity and customer satisfaction, and a strategy of radical improvement most explained relative technological competitiveness. Furthermore, motivation and resource-based theories can explain the link between a firm's clarity of strategy and its success. Clear and explicit goals on NPD, sustainable development, e-Commerce, or innovation in general motivate people and channel resources in organizations to achieve those goals.

This explanation is supported by industry examples, such as 3M Corporation's clear objective of having 25 percent of its sales generated by products introduced in the most recent five years (Shrivastava, 1995), a goal that has now been increased. In summary, a clear innovative strategy that fits in with the overall organization's strategy, and the clear understanding or definition of various aspects of innovation (such as NPD or the new technology adoption objectives of the organization) should be related to innovation capability.

Organizational Structure

An organization's structure is "the formal setup of a business corporation's value chain components in terms of work flow, communication channels, and hierarchy" (Wheelen & Hunger, 2002, p. G-5). Past research has examined the relationship between innovation and organizational structure by considering the degree of specialization, functional differentiation, centralization, formalization (i.e., the level of reliance on rules and procedures), as well as other aspects of structure.

The results of past research indicate that the relationships between innovation and various aspects of structure differ depending on how innovation is measured. For example, Damanpour (1991) found a positive link between innovation and specialization, functional differentiation and the ratio of managers to the number of employees in the organization (which he coined "administrative intensity").

Damanpour (1991) defined specialization as the different specialties (typically identified by the number of occupational types or job titles) found in organizations. In turn, Damanpour (1991) explained that functional differentiation "represents the extent to which the organization is divided into different units" (p. 589).

In contrast, Subramanian and Nilakanta (1996) found a negative association between specialization and administrative innovations (p. 639). Administrative innovations "involve organizational structure and administrative processes; they are related to the basic work activities of an organization and are more directly related to its management" (Damanpour, 1991, pp. 560–561). Administrative innovations may "constitute the introduction of a new management system, administrative process, or staff development program" (Subramanian & Nilakanta, 1996, p. 637). With regard to centralization, Damanpour (1991) found that innovation was negatively related to the centralization of decision-making autonomy. However, Subramanian and Nilakanta (1996), for instance, found that centralization was negatively associated to the time of technical innovation adoptions, but positively associated to the time of the administrative adoptions.

Furthermore, Damanpour (1991) found that the degree to which organizations had established rules and procedures did not appear to be related to innovation. However, innovation has been linked to flatter and more flexible reporting structures rather than to more hierarchical and inflexible ones (e.g., Ahmed, 1998). In summary, past research indicates that the relationships between some measures of structure, such as centralization and specialization, and innovativeness are moderated by the type of innovation. In addition, the centralization of decision-making and hierarchies with many levels of management might hinder some aspects of innovation, but the degree of formalization, specialization, functional differentiation, and the ratio of managers to the number of employees might not.

Type of Organization

The predictors of innovation appear to depend on the type of organization. For instance, Damanpour (1991) found that the standardization of work processes facilitated innovation in manufacturing, but hindered it in service organizations. In contrast, vertical differentiation (or the number of hierarchical levels in the organization) hindered innovation in manufacturing, but not in service organizations. Similarly, Atuahene-Gima (1996) found that there were some differences in the factors related to innovation success in Australian services and manufacturing firms. For example, Atuahene-Gima (1996) found that the newness of the innovation to customers has a negative impact on the success of new services, but not on the success of new products. Therefore, based on past research, empirically distinguishable theories of innovation need to be developed for the manufacturing and the service industries, as suggested by Damanpour (1991, p. 575).

Slack Resources

Past research found a relationship between "slack resources" and innovation (e.g., Ahmed, 1998; Damanpour, 1991) in general, and NPD (Cooper & Kleinschmidt, 1996) and technology innovation (Caldeira & Ward, 2003) in particular. Slack resources reflect "the resources an organization has beyond what it minimally requires to maintain operations" (Damanpour, 1991, p. 589).

Slack resources are typically measured as human and financial resource slack. For instance, based on a research study of 500 Australian small to medium enterprises, Terziovski (2003) found that the allocation of resources to maintain formal support systems such as job rotation, training and communication linkages was one of the best indicators of innovation performance SMEs. Having slack resources is likely to be a contributing capability factor of innovative organizations.

Culture and Climate

A climate of creativity, welcoming of change, tolerant of risk and failure, and democratic, is more likely to be linked to innovation than a culture that discourages creativity, does not welcome change, is not tolerant of risk, and is autocratic (e.g., Angle, 1989; Ahmed, 1998; Roberts, 1996). For example, Terziovski (2003) found that innovation performance was positively related to employees being able to use failures as opportunities to learn.

However, Angle (1989) cautioned that creativity is not innovation and proposed, "Innovation effectiveness will be positively related to the extent to which the organization is able to integrate creative personalities into the organizational mainstream" (p. 152). In reality, innovative organizations usually encourage, mentor and support idea champions (Angle, 1989; Ahmed, 1998; Roberts, 1996; Shrivastava, 1995) and model innovative behavior (Angle, 1989). This approach to people leads to a climate of creativity and innovation. Thus, a democratic, risk-tolerant and people-oriented culture appears to characterize organizations capable of innovation.

Communication

Internal and external communication have been found to be related to innovation (e.g., Ahmed, 1998; Damanpour, 1991; Gleadle, 1999). External communication encompasses knowledge of the environment outside the organization, interaction with customers and suppliers, and the involvement of employees with external professional activities and networks. Internal communication involves, for example, the clear communication of the firm's NPD approach (e.g., Cooper & Kleinschmidt, 1996), e-Commerce usage and sustainable development (e.g., Azzone & Noci, 1998) strategies, which is critical to the firm's performance in each of these areas.

In addition, Angle (1989) proposed that innovation capabilities could be enhanced by the frequency of communication that occurs horizontally (across functions) and vertically (across managerial levels) in the organization. Consequent to the above discussion, communication appears to be a capability of innovative organizations. This is because effective internal and external communication enables information to be shared by all stakeholders in the innovation process.

Social Structures

Interpersonal relationships that facilitate innovation in organizations are part of what we call "entrepreneurial networks" in this literature review. Dougherty and Bowman (1996) explained entrepreneurial network as people's "personal influence to get support from others and to champion the new product; they reached across formal boundaries informally to gather resources; they drew on their experience and their understanding of the firm's history to connect with senior managers" (pp. 31–32). Kanter (1984) referred to the ability to persuade people to invest time and resources in risky propositions as the "power skills" that managers must have to lead change. In support, past research has shown that interpersonal relationships and the distribution of power in organizations are key determinants of innovation success (Caldeira & Ward, 2003; Dougherty & Hardy, 1996).

However, these entrepreneurial networks can be broken when an organization downsizes. The effect of downsizing on innovation effectiveness is an important one because, as Dougherty and Bowman (1996) explain, "even at the firm with the most highly developed new product system ... innovators relied heavily on personal influence and networking" (p. 34). To preserve the entrepreneurial network, Dougherty and Bowman (1996) recommend organizations minimize the anti-innovation effects of downsizing by, for example, retaining the top management sponsor and operating level champion of particular innovative projects. The social networks of a firm are a contributing factor to its innovation capability.

People and HRM

Angle (1989) looked at organizational innovation from a psychological perspective and proposed that organizational innovation was a combination of the members' personal attributes and the organizational context. In addition, he proposed that organizational innovation only occurs if organizations provide a context that motivates and enables innovation (p. 139). Therefore, we review the literature on managerial skills, leadership, knowledge, teams, incentives and rewards.

• Management and leadership

Leaders and managers influence what is done in organizations, as well as how things are done. Therefore, it is important to understand the relationship between various aspects of leadership and management behavior, and organizational innovation. Barker and Mueller (2002) found that between 11 percent and 14 percent of the variance in relative R&D spending was predicted by CEO characteristics. Specifically, R&D spending was greater in firms with younger CEOs, who had greater wealth invested in firm stock, had significant career experience in marketing and/or engineering or R&D, and had advanced science-related degrees. In addition, a management philosophy that supports innovation and people has been found to characterize the highly innovative company (e.g., Ahmed, 1998; Damanpour, 1991; Roffe, 1999).

Damanpour (1991, pp. 569, 589) found that the extent to which the managers or the members of the dominant coalition in the organization favored change predicted innovation. Atuahene-Gima (1996) also found that management support (and teamwork) was related to innovation success for both services and manufacturing firms in Australia. Similarly, Caldeira and Ward (2003) found that the attitudes and support of top management towards information systems or information technology (IS/IT) adoption and use in small to medium manufacturing companies in Portugal was one of the two determinant factors in IS/IT success. The second one was IS/IT people skills and knowledge. Senior management commitment is also a critical factor in NPD (Cooper & Kleinschmidt, 1996; Mabert et al., 1992), in e-Commerce (e.g., Chan & Swatman, 2000) and in sustainable development initiatives (e.g., Dunphy et al., 2003; Azzone & Noci, 1998). In addition, senior management accountability for the success of innovative endeavors is linked to the firm's performance (e.g., Cooper & Kleinschmidt, 1996). Further, while US companies were found to successfully scrimp in new product development by using project managers with low technical or low human resources skills, their success rates were lowest when project managers had low leadership skills (Souder et al., 1998). Souder et al. (1998) described leadership skills in their study as "the ability to inspire others in achieving the project's goals" (p. 53). In summary, CEO characteristics, leadership skills, and a management philosophy that supports innovation and people can positively influence innovation capability in organizations.

- *Knowledge*

Having people with the technical and professional knowledge, keeping knowledge in-house and being able to leverage from it by sharing it appear to be important factors to innovation. Damanpour (1991) found that the technical and professional knowledge of the people in the organization were positively linked to innovation. "Technical knowledge" reflects the organization's technical resources and technical potential, and "professional knowledge" reflects both education and experience (Damanpour, 1991). In support, Caldeira and Ward (2003) found that the quality of the IS/IT people (their skills and knowledge) was a determinant factor in IS/IT success in SMEs in manufacturing, in Portugal. Knowledge can be captured, shared and enhanced in many ways. Some companies have introduced expert systems to capture and share organizational knowledge (e.g., Ahmed, 1998; Mort & Knapp, 1999), or used internal (intranet) and external (internet) communication systems to assist with the sharing and searching of new ideas (Roffe, 1999).

In addition, companies have been able to analyze their current learning and knowledge generation processes within product innovation, and then have improved those processes by using the CIMA Behavioral Model (Boer *et al.*, 2001). Yet, Souder *et al.* (1998) found that US companies were likely to still achieve acceptable (38 percent) percentages of success in new product development when the technical skills of the project manager were low. "Acceptable" percentages were described by Souder *et al.* (1998) as success rates not statistically lower than those obtained by chance (p. 54). Nevertheless, Souder *et al.* (1998) found that new product success rates were high (65 percent or above) for project managers with high technical, HR, leadership and marketing skills.

In addition, Roffe (1999) concluded from his literature review on creativity and innovation that core training activities to build the organization's intellectual capital are central to creating capabilities for improved performance. At a national level, the level of education of a population was found to be related to the country's capacity to innovate (Porter & Stern, 1999). Therefore, the technical and professional knowledge of managers and employees in the organization are important to innovation capability, including new product success.

- *Teams*

Much of the past research indicates that the use of teams is beneficial to productivity, organizational performance, accelerated NPD and innovation. For instance, amongst other positive outcomes, teaming and physical collocation of teams has been found to reduce product design, lead time and development, and to achieve productivity gains (e.g., Mabert *et al.*, 1992; Pawar & Sharifi, 1997; Sohal *et al.*, 2003). Collocation was defined by Rafii (1995; cited in Pawar & Sharifi, 1997) as

> ... physical proximity of various individuals, teams, functional areas and organizational sub-units involved in the development of particular product and process (p. 285).

Further, Atuahene-Gima (1996) found that teamwork (and management support) was the second most important factor of innovation success for services and manufacturing firms in Australia. In addition, high-quality development teams and cross-functional teams appear to be two of the factors that are critical in new product development (e.g., Ahmed, 1998; Cooper & Kleinschmidt, 1996; Mabert *et al.*, 1992).

For example, Ahmed (1998) found that highly innovative organizations deliberately staff design and development teams with individuals from different functional departments. In turn, high-quality teams were characterized by having a dedicated project team leader, having frequent team meetings, and by handling decisions quickly and efficiently (Mabert *et al.*, 1992).

In addition to high-quality and cross-functional teams, group cohesiveness may also be positively associated with innovative effectiveness, providing a climate of trust prevails among the members of the team (Angle, 1989, p. 160). Nevertheless, teaming can also produce some undesirable outcomes. For instance, teams may be constrained by "groupthink" (Samson & Daft, 2003), and co-location may exert disparate effects on physical and virtual teams (Pawar & Sharifi, 1997). However, overall, past research indicates that the use of teams in organizations can generally promote innovation capability.

- *Incentives and rewards*

Incentives and rewards are used to motivate people and channel efforts towards the achievement of organizational and personal goals. Angle (1989) concluded from his review of the literature that individual rewards tend to increase idea generation and radical innovations, while group rewards tend to increase innovation implementation and incremental innovations. In addition, innovative behavior is a function of the intrinsic and extrinsic rewards provided by the organization (Angle, 1989; Roffe, 1999). Both Angle and Roffe conclude that recognition is likely to be a more effective extrinsic reward than money. Therefore, it is concluded that incentives and rewards can be used to harness innovation capability in organizations.

The Management of Technology

An OECD *Policy Brief* (2000) noted that firms' investment in information and communication technologies (ICT) has increased in recent years, and that the services sector is the main purchaser of ICT equipment. The *Policy Brief* concluded that ICT enable many changes in the economy and in the innovation process, and help make other economic sectors more innovative (p. 7). However, past research indicates that the successful adoption and the use of new technology depend on organizational and environmental characteristics, as well as on whether the technology change is radical or incremental. For example, some of the specific organizational capabilities needed for incumbent organizations to survive radical technological changes appear to be the legitimization and institutionalization of autonomous action, the gestation period of the technology innovation, the history of the incumbent firm, and the organization's resource slack (Hill & Rothaermel, 2003).

In addition, for small- to medium-size manufacturing firms, the quality of external technical expertise and services available and the cooperation with external suppliers are some of the factors found to be related to the success of new technology adoption and use (Caldeira & Ward, 2003). Technology also plays a role in successful NPD, e-Commerce initiatives and innovation involving sustainable development. For example, it appears that technological synergy, such as the joint use of computer-aided drafting (CAD) and manufacturing (CAM), may accelerate NPD (Cooper & Kleinschmidt, 1996; Mabert et al., 1992).

Further, new technology adoption and use is also important in e-Commerce (Gebauer & Scharl, 2003) and sustainable development (Dunphy et al., 2003; Shrivastava, 1995) initiatives. With regard to the internet, the current view is that the internet is not only a necessary tool to conduct business in the current environment, but it can also be an enabling technology that allows organizations to gain a competitive advantage (Porter, 2001). However, Porter (2001) warns that the internet should be used to complement and enhance the organization's current capabilities, rather than as a separate identity. In sum, research indicates that adoption of technology can enable innovation (e.g., Mort & Knapp, 1999; Terziovski, 2001). However, the type of technological change and the organization's characteristics determine the effective adoption and use of technology in achieving organizational innovation.

Market Knowledge

The importance of market knowledge has been demonstrated in the cases of winning strategies (Wheelen & Hunger, 2002) in general, and considered to be an important factor in the cases of NPD (e.g., Cooper, 1985; von Hippel et al., 1999) and of technological innovation (e.g., Delaplace & Kabouya, 2001; Sharfman et al., 2000) in particular. For example, von Hippel et al. (1999) described the "lead user" process as a means to successfully and systematically innovate. This process is based on development teams identifying and working closely with "companies or people that have already developed elements of commercially attractive breakthroughs" (p. 48).

However, while Terziovski (2003) did not find support for the hypothesis that formal monitoring of developments in new technology was positively related to increased business excellence for small and medium enterprises (SMEs) in Australian manufacturing, he found that such monitoring of developments in new technologies was related to an increase in the number of new ideas adopted, the success of new products, and increased market opportunities.

Generally, environmental scanning and knowledge of the market is necessary to formulate strategy (Cooper, 1985; Wheelen & Hunger, 2002). Therefore, without market knowledge, companies might not be able to formulate a strategy with regard to innovation, NPD, sustainable development, or technology and e-Commerce that will lead to outperforming their competitors. Market knowledge appears to be a necessary component of effective innovation capability.

2.6 Integrating e-Commerce, Sustainable Development, NPD and Innovation Capability

In this section we review the roles of e-Commerce, sustainable development and NPD in relation to an organization's innovation capability. Fig. 2.1 shows the relationship between these constructs.

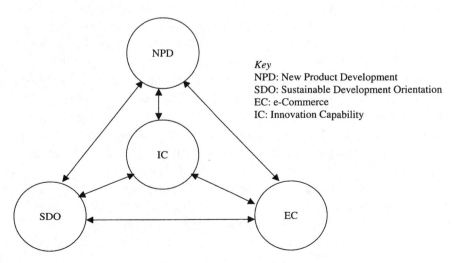

Key
NPD: New Product Development
SDO: Sustainable Development Orientation
EC: e-Commerce
IC: Innovation Capability

Fig. 2.1 Integrated innovation capability model.

e-Commerce

The authors did not find empirical literature on e-Commerce and its role in organizational innovation capability. Most of the literature on e-Commerce focuses on the implementation of technologies prior to the advent of the internet, such as EDI (Electronic Data Interchange) (e.g., Ngai & Wat, 2002). Those that compared the adoption or intended use of internet-based technologies for e-Commerce purposes with the level of adoption of prior technologies, found evidence that emerging internet-based technologies are easier to adopt by organizations of all sizes than EDI (e.g., Chan & Swatman, 2000; Porter, 2001). This is chiefly due to the low cost of internet-based technologies.

Past research on e-Commerce and the use of internet-based technologies also found that the first step to using the internet for e-Commerce purposes has been the setting up of a website that provides information about the company (Chan & Swatman, 2000; Porter, 2001). In addition, large firms are more likely than small ones to have e-Mail and/or access to the internet, to have their own website, to sell online (but no more than 21 percent appear to do so), and to buy online (e-Procurement) (Konings & Roodhooft, 2002).

In addition, recent literature on e-Commerce indicates that key success factors in the implementation of e-Business strategies overlap with key success factors in organizational innovation in general (e.g., Chan & Swatman, 2000; Chang *et al.*, 2002; Konings & Roodhooft, 2002; Phan, 2002). Some of these factors are support from top management; the involvement in the implementation process not only of top management, but also of the managers most responsible for the implementation; the technical and other support of trading partners to reduce resistance to change; proper planning; and good (e-Business) management strategies, such as having clear goals and well-defined scopes (e.g., Chan & Swatman, 2000; Phan, 2002).

Furthermore, based on a case study of Intel, Phan (2002) concluded that other key success factors include a distinctive strategic position, use of e-Business to complement traditional ways of competing, focus on quality connections, worldwide support and customer training, a robust e-Business architecture, and use of security protections to build customer trust (pp. 587–589). Moreover, corporate e-Commerce strategy appears to be related to the successful implementation of e-Commerce initiatives.

From content analyses of the letters of CEOs that appeared in the year 2000 annual reports of large organizations, Chang *et al.* (2002) found a positive relationship between a firm's perception of the importance of e-Commerce and the firm's performance. Overall, firms appear to benefit from having e-Commerce integrated with corporate strategy. If not well-managed, some of the above factors can inhibit the adoption of e-Commerce. Debreceny *et al.*'s (2002) study, conducted in Singapore, identified seven clusters of inhibiting factors.

Cluster 1 reflected issues of brand building, primarily with regard to business-to-consumer (B2C). Cluster 2 had a strong management skills and attitudes orientation, whereby the attitudes and skills of senior managers were seen and described by the participants as significant inhibitors to the adoption of e-Commerce. Cluster 3 was characterized by the issue of consumer trust, whereby the lack of trust in B2C is still a problem. Clusters 4 to 7 covered a range of attitudes of participants towards e-Commerce. Furthermore, in the scarce empirical research on the effect of e-Business on corporate performance, we found differential effects by organization size and industry type.

For example, Konings and Roodhooft (2002) found in their sample of 836 Belgian firms that e-Business had no effect on the productivity of small firms, but had a positive effect on the productivity of large ones. Konings and Roodhooft (2002) also concluded from the results of their study that the use of e-Business is limited, but substantially higher among large than small firms. Chan and Swatman's (2000) study of the BHP Steel (Australia) experience suggests that technical issues (e.g., technical complexity and compatibility) are more important at the early stages, and management (e.g., management commitment) and business (e.g., trading partner relationships) issues at the late stages of the e-Commerce implementation. In sum, the authors found no studies that specifically examine the relationships between e-Commerce and innovation capability in organizations. This is congruent with Ngai and Wat's (2002) literature review findings and classification scheme for e-Commerce research. However, we know that communication intensity and effectiveness are underpinning factors that drive innovation capability, and e-Commerce is a new and powerful form of such technology.

2.7 Sustainable Development (SD)

Researchers agree that there is no one set of practices that comprise SD practices and apply to all enterprises across all industries (Goldsmith & Samson, 2002; Hunt & Auster, 1990). This is partly because the appropriate mix of practices that maximizes the SD needs and the strategic objectives for one organization may not be appropriate for another. Examples of SD practices are waste minimization, recoverable manufacturing and supplier protocols. Cerin and Karlson (2002) showed that the emission costs per net sales were generally higher for manufacturing companies (e.g., home appliance, IT and telecom, vehicle manufacturer, chemistry, electric power and mining) than for service products (e.g., broadcasting and mobile telecom provider), but exceptions existed (e.g., airline and road transport). Therefore, sustainable development is of great importance to all industries, but possibly most directly to the manufacturing industry.

Goldsmith and Samson (2002) propose a model of the relationships between sustainable development practices and business success that takes into account differences between industries. Their propositions also reflect the influence of an industry sustainable development orientation (SDO) on the enterprise's SDO. They propose that enterprises with higher SDO are more likely to be successful in the long-term, but not necessarily in the short-term, and to be leaders in their industry SDO than those with lower SDO. An organization's SDO can be seen as the degree to which the organization meets not only its economic goals, but also its environmental ones. In turn, sustainability can be defined as "meeting our current needs without jeopardizing the ability of future generations to meet theirs" (Shrivastava, 1995, p. 184).

Other researchers have proposed models that explain where organizations are at with regard to their commitment to environmental issues (e.g., Dunphy *et al.*, 2003; Hart, 1997; Hunt & Auster, 1990). For example, Hunt and Auster (1990) found that organizations can be in one of five distinct stages of environmental program development. The stages range from the "beginner" to the "pro-activist", and are assessed on a number of criteria such as the degree to which the program reduces environmental risk and the level of top management support and involvement (p. 9).

They found that a surprisingly large number of corporations in their survey fell in the "beginner" and "fire fighter" stage and very few in the "pro-activist" stage. In support, Hart (1997) stated that few companies have incorporated sustainable development in their business strategy, and that most firms are still at the first stage of pollution control rather than prevention; yet, corporations need to become more environmentally conscious if they want to remain competitive in the long-term, for two reasons.

Firstly, the view is that consumers will increasingly demand environmentally friendly products (e.g., Cormier *et al.*, 1993; Shrivastava, 1995). Secondly, integrating environmental technologies into strategic management should offer advantages such as reduced operating costs (e.g., reduced waste and increased energy conservation), enhanced revenue (e.g., through increased customer loyalty), closer ties with suppliers, improved quality programs, competitive advantage through greater profitability and the creation of unique strategies, reduced liabilities by better managing of environmental risks, and enhanced public image, amongst others (Dunphy *et al.*, 2003; Shrivastava, 1995).

Environmental technologies were defined by Shrivastava (1995) as "production equipment, methods and procedures, product designs, and product delivery mechanisms that conserve energy and natural resources, minimize environmental load of human activities, and protect the natural environment" (p. 185). Shrivastava (1995) used the 3M Corporation to illustrate how an environmental focus when using technologies saved money for 3M, enhanced its public image and product appeal, and ultimately set new standards of environmental performance and, thus, reshaped the industry's competitive dynamics. Therefore, environmentally conscious innovation is believed to not only reduce the costs of production, but also to improve a firm's competitive advantage through enhanced financial performance and public image.

The extant body of knowledge indicates that many of the factors that foster the successful introduction of environmental innovations are similar to factors related to innovation in general. For example, top management commitment to innovation-based environmental strategies, champions with company-wide support, a "learning" culture supported by training, worker empowerment, and external factors such as government regulation can trigger the successful implementation of environment-driven change (e.g., Azzone & Noci, 1998; Dunphy *et al.*, 2003; Roberts, 1996).

Furthermore, the results of Azzone and Noci's (1998) study of a small sample of 15 organizations suggest that cooperative relationships with various stakeholders (supplier, customers and even competitors), business strategies that encompass and support "green" innovation, and communication are a few of the changes necessary for successful implementation of innovation-based environmental strategies. In addition, an enterprise's SDO and performance appear to be linked to internal and external communication; motivation; training and management of people in the organization; the firm's strategy, structure and culture; and its financial, technical and social resources (Dunphy *et al.*, 2003; Goldsmith & Samson, 2002; Hunt & Auster, 1990).

Sharfman *et al.* (2000) also found some support for the propositions that economic incentives for environmentally conscious innovation development, the costs of pollution, and the level of flexibility the firm has in meeting its environmental objectives increase a firm's likelihood to engage in environmentally conscious (technological and process) innovation. However, Sharfman *et al.* (2000) found little support for predictions that restrictive decision rules and political behavior would decrease the likelihood of firms engaging in environmental innovation.

From the literature reviewed, it is clear that although many of the organizations may have the competence to optimize the environmental performance of their products, they are not motivated under the current policy instruments to do so (Cerin & Karlson, 2002). Roberts (1996) pointed to champions that had wide company support and to a supportive regulatory system as key factors in the success of one of the companies studied. Barriers to developing a SDO appear to be a conservative and autocratic leadership, a tight financial position, lack of information on the environmental benefits of projects, and a short-term approach to assessing the feasibility of innovation projects (e.g., Roberts, 1996).

2.8 Future Research Agenda

As a result of the extant body of research, we now have a general understanding of what organizations need to do to successfully manage innovation. However, most of the research to date has focused on the general notions of innovation capability such as leadership, organizational knowledge and skills, culture, human resource management, structure and strategy, and industry factors (e.g., Ahmed, 1998; Damanpour, 1991; Cooper & Kleinschmidt, 1996; Roffe, 1999). The authors found no studies that deeply examine the relationships between e-Commerce and innovation capability in organizations. The little research that exists on e-Commerce used single case studies, focused primarily on large firms, and used high-level financial measures that might not have revealed the true value of e-Commerce initiatives (e.g., Chan & Swatman, 2000; Konings & Roodhooft, 2002; Phan, 2002).

Past research also does not explain how innovation capability manifests in sustainable development innovations or in new product development. Much of the information on sustainable development to date stems from non-empirical work (e.g., Cerin & Karlson, 2002; Goldsmith & Samson, 2002; Rennings, 2000) and descriptive analyses of small samples of data (e.g., Azzone & Noci, 1998; Sharfman *et al.*, 2000) and case studies (e.g., Roberts, 1996). Some authors did not explain the analyses or samples used in their studies (e.g., Hunt & Auster, 1990).

Overall, the literature on "green" innovation lacks the rigor of multivariate empirical analyses that use triangulated methodology approaches, including multiple sources of data (e.g., Azzone & Noci, 1998; Hunt & Auster, 1990; Roberts, 1996; Sharfman et al., 2000). Similarly, most of the studies on NPD have methodology limitations, such as small sample sizes, use of secondary data, and qualitative comparisons of case studies on NPD (e.g., Gleadle, 1999; Firth & Narayanan, 1996; Mabert et al., 1992). For example, small sample sizes and case studies hinder the generalizability of the results of past research. Therefore, much more still needs to be done to fully understand the roles of e-Commerce, sustainable development and accelerated NPD in the innovation capability of organizations.

Further, organizational innovations can be categorized in various ways, but researchers have usually focused on one category. Categories used in past research include technical versus administrative innovations, radical versus incremental innovations, and initial versus implementation stages of innovation (e.g., Subramanian & Nilakanta, 1996). Wolfe (1994) also identified the focus on a single organization as a barrier to accumulating knowledge in innovation research. In addition, most studies did not examine the relative importance of the factors related to innovation.

Past research used chiefly qualitative data analyses based on case studies (e.g., Dougherty & Hardy, 1996; Roberts, 1996), descriptive statistics and simple inferential data analyses methods (e.g., Acs & Audretsch, 1988), or only considered one block of variables instead of their relative importance to the outcome variable (e.g., Theyel, 2000). As a result, there is a gap in our knowledge regarding the relative importance of the various external and internal factors to the innovation capability of organizations, because of the methodology used to date.

An understanding of the relative contribution of various factors to innovation capability is needed, so that organizations can make the right decisions on how to develop and exploit innovation capability to achieve organizational goals. For instance, are human resource management (HRM) practices and capabilities more strongly related to innovation performance than non-HRM factors? Based on scant research on the relative importance of these two sets of factors, the HRM factors appear to be more important than the non-HRM factors to an organization's capability to benefit from ISO9000, TQM and innovation initiatives (e.g., Samson & Terziovski, 1999).

It is important to note that the relative importance of the facilitators and inhibitors of innovation may change as the innovation moves from the initial stage of design, to the production, implementation and commercialization stages. This needs to be examined empirically. Furthermore, most of the empirical research on innovation conducted to date has focused on a limited number of factors and their impact on one dimension of performance (such as customer service or the number of successful new or modified products to market). Although the direct impact of any one factor on innovation is important, it is the combined effect of several factors that can produce powerful results (e.g., Laursen & Foss, 2003; Mort & Knapp, 1999).

For example, Laursen and Foss (2003) found support for the hypothesis of complementarities between certain HRM practices. Laursen and Foss (2003) included nine discrete variables pertaining to new HRM practices in their study. The nine variables were combined into two HRM systems. Both HRM systems more strongly explained innovation performance than each of the individual HRM practices. Past research also indicates that the relationships between some predictors and each dimension of innovativeness may vary (Subramanian & Nilakanta, 1996).

In addition, despite the valuable work reviewed (e.g., Dunphy *et al.*, 2003; von Hippel *et al.*, 1999; Mabert *et al.*, 1992; Sharfman *et al.*, 2000; Terziovski, 2003), there is insufficient empirical research on how formal and informal partnerships foster innovation capability in organizations. Furthermore, in order to fully understand the notion of innovation capability, future research would benefit from international comparisons.

There are some studies in innovation that use data from several countries (e.g., Hagedoorn & Cloodt, 2003; Konings & Roodhooft, 2002). In particular, Porter and Stern (1999) compared the national innovation capacity of 17 countries over a period of 25 years. However, to the authors' knowledge, there are no studies that systematically and broadly compare the characteristics of innovation capability at the organization level, across countries. Therefore, future research can benefit from an integrated and comprehensive examination of the relationships between groups of internal and external factors, and a composite measure of innovation capability in organizations. The roles of e-Commerce, sustainable development orientation and NPD should be included in this research approach. This approach will enable researchers and practitioners to obtain the answers to the following research questions:

(1) What are the performance effects of innovation capability?
(2) How does innovation capability manifest in the new economy (e-Business) firms?
(3) How does innovation capability manifest in sustainable development innovations?
(4) How does innovation capability manifest in NPD processes?
(5) How do formal and informal collaborative partnerships foster innovation capability?
(6) What are the specific and different characteristics of innovation capability for the manufacturing and services industries?
(7) What are the international differences in the nature of organizational innovation capability?

In order to address the above research questions, we suggest that research into innovation capability in the next several years can benefit from the following recommendations. First, researchers should conceptualize innovation capability as a multi-dimensional construct (Cooper *et al.*, 1998). In support, Hagedoorn and Cloodt (2003) found it advantageous to use a four-item construct to measure innovative performance in high-tech industries. Multi-item dependent variables are better than single-item ones at explaining complex phenomenon such as that of sustainable innovation capability and its effects on organizational performance.

2.9 Synthesis of the Discussion

The aims of this chapter were to identify from existing literature the characteristics of innovation capability, and to set the research agenda for the next several years in innovation capability. As a result of the literature reviewed, we formulated seven research questions and put forward seven broad suggestions for future research. Overall, past research in innovation management has provided information on the factors characteristic of highly innovative versus less innovative companies, which in turn has enabled the development of models of innovation. For example, Lawson and Samson (2001) proposed an "innovation capability" construct with seven elements, namely, "vision and strategy, harnessing the competence base, organizational intelligence, creativity and idea management, organizational structures and systems, culture and climate, and management of technology" (pp. 377, 387–395).

From the literature reviewed, it appears that many factors (or constructs), singly and combined, can positively impact on an organization's capability to continuously innovate on its performance. For example, with regard to external factors, government regulation appears to be a critical factor for the successful pursuit of environmentally sustainable or e-Commerce opportunities. In addition, the set of sustainable development (SD) practices and the firm's approach to SD are likely to depend on the industry the organization is in. In turn, internal factors such as management style, technology, and the firm's financial position are likely to influence the firm's approach to sustainable development.

Many of the external and internal factors linked to the more general notion of innovation appear to be also linked to successful NPD, e-Commerce and sustainable development initiatives. This indicates that there is a core of capabilities that apply to the notion of continuous innovation capability in general. Some of these core capabilities appear to be top management's commitment, characteristics and ability to lead, cultures tolerant of risk, clear strategies, comprehensive internal and external communications, cooperative internal relationships and external partnerships, cross-functional development teams, skilled people, and slack human and financial resources. Further, many of the relationships appear to be multi-directional.

For example, the regulatory environment may influence an organization's approach to sustainable development, but organizations

that are leaders in sustainable development practices can also influence government regulation. Based on Ahmed's (1998) study, hard
capabilities reflect methods, technology, organizational systems, organizational infrastructures, and non-people and people resources, characteristic of innovative companies. In turn, soft capabilities refer to
the proper and effective management and leadership of the organization's hard capabilities.

Future models of innovation capability need to empirically examine this balance using different methods (e.g., survey and case study
approaches) and large, multinational samples. In addition, future
models of innovation capability will benefit from an acknowledgment
that innovation capability is likely to differ between the service and
manufacturing industries. Finally, future models need to integrate
general notions of innovation capability with organizational capability
in e-Commerce, sustainable development and NPD. Only then will
we be able to gain a comprehensive understanding of what constitutes
continuous innovation capability in organizations across different
industries and countries.

2.10 Conclusion

In this chapter we addressed the question what constitutes innovation
capability in organizations, and how can it be developed and exploited?
Based on our extensive literature review, we conclude that there are
no clear, agreed guidelines for creating innovation-driven organizations. Numerous studies have attempted to isolate the important variables facilitating innovation outcomes (Damanpour, 1991). However,
there is still much we do not know about how firms can innovate
faster and better. We do know that effective innovation requires the
construction of an overarching framework of factors conducive to creativity (Amabile *et al.*, 1996; Kanter, 1989).

The absence of such frameworks may lead to a conservative and
ineffective innovation culture. We further conclude, based on our
analysis, that innovation capability provides the potential for effective
innovation. It is not a simple nor single-factored concept, as it involves
many aspects of management, leadership and technical aspects, as well

as strategic resource allocation, market knowledge, organizational incentives, etc. Overall, it appears that a balance between hard and soft capabilities for innovation is necessary for innovation to be successful and sustainable (e.g., Ahmed, 1998; Roberts, 1996). In particular, we conclude that the integration of e-Commerce, sustainable development and new product development with the general notions of innovation capability is a key imperative for the creation of highly innovative organizations.

Review Questions

(1) Discuss the internal and external factors that have a positive impact on an organization's capability to continuously innovate.
(2) Explain why HRM factors appear to be more important than non-HRM factors to an organization's innovation capability.

Chapter 3

Strategic Shift from Product Orientation to Innovative Solutions Capability in the German Biotechnology Industry: Sartorius AG

Milé Terziovski and B. Sebastian Reiche

3.1 Introduction

The biotechnology industry is still considered an early stage industry that has maintained its link to scientific knowledge sources. While this linkage ensures a high knowledge and research intensity, the market dynamics make biotechnology firms highly dependent upon the development and commercialization of innovation. As a result, these organizations serve as valuable units of analysis for studying determinants and outcomes of innovation processes (Gittelman & Kogut, 2003; Rzakhanov, 2004).

Several researchers have conceptualized innovation management as a type of dynamic organizational capability (Teece & Pisano, 1994; Lawson & Samson, 2001). A key characteristic of dynamic organizational capabilities is their capacity to provide inimitable combinations of resources that cut across corporate functions and ensure a sustainable competitive advantage (Teece & Pisano, 1994; Eisenhardt & Martin, 2000). Consequently, the main challenge for researchers and practitioners alike is to identify relevant components of innovation capabilities that serve as the main drivers for corporate innovation and their application in different organizational contexts.

Adopting a single case study approach (Eisenhardt, 1989; Yin, 2003), this study sheds light on prevalent innovation capabilities in the biotechnology industry, focusing particularly on Germany-based Sartorius AG. A special focus lies on the role of e-Commerce, sustainable development and new product development in facilitating the innovation process (Metz *et al.*, 2004). Moreover, the company's innovation strategy and knowledge-sharing culture are examined. In doing so, the study presents an analysis of how the company transformed itself from a traditional product-oriented company to a total solution provider in order to maintain its leading market position in the field of biotechnology and mechatronics.

The case study is based on an in-depth interview with two senior managers at the biotechnology division of Sartorius AG, the Senior Vice-President of the business area Bioprocess (BP) and the Head of the business unit Purification Technologies (PT), respectively. While the former has been working for the company for over 21 years and is one of the longest employed, the latter, coming from a big German chemical company, has only recently joined the company in order to provide process know-how with regard to establishing the new business unit, PT. A series of company information like company brochures, the company website and annual reports were scanned for relevant information and served as a means of data triangulation (Miles & Huberman, 1994).

The text is divided into six sections. First, the company background is described. Second, the study examines the company's strategic position with regard to innovation and subsequently investigates elements of its capacity to absorb new knowledge and integrate it into the organizational knowledge base. The main section then analyzes major corporate innovation capabilities with a special focus on knowledge management, sustainable development, e-Commerce and new product development. This section is followed by remarks on organizational performance outcomes. The text concludes by highlighting organizational implications for academics as well as managers and summarizing the main findings.

3.2 Company Background

Sartorius AG is an internationally leading laboratory and process technology supplier covering the segments of biotechnology and mechatronics. In 2003, the technology group earned sales revenues of 442.3 million euros. The company is based in Goettingen, Germany. It was founded in 1870 and finally became a public company listed on the German stock market in 1990. It currently employs just over 3,660 people. Sartorius maintains its own production facilities in Europe, Asia and America as well as sales subsidiaries and local commercial agencies in more than 110 countries.

Its biotechnology segment focuses on filtration and separation products, bioreactors and proteomics. It embraces five business areas — Biolab, Bioprocess, Biosystems, Food & Beverage and Environmental Technology, each subdivided into different business units. The mechatronics segment particularly manufactures equipment and systems featuring weighing, measurement and automation technology for laboratory and industrial applications as well as bearings, especially hydrodynamic versions. This division consists of four business areas: Lab Instruments, Process Weighing & Control, Service and Hydrodynamic Bearings.

Sartorius' key customers are from the pharmaceutical, chemical and food and beverage industries and from numerous research and educational institutes of the public sector. The biotechnology division mainly concentrates on the biopharmaceutical industry with approximately 70 percent of its customers being biotech companies. In line with the current strategic scope of the company, its biotech drugs now derive from biotech processes rather than, like in the past, from chemical processes. Certain external, industry-specific constraints exist for Sartorius' innovation activities. As the Head of PT noted:

> [t]he problem is that we are serving a very conservative industry that doesn't like change. That is even more the case the more complex the product. If you have a certain technology available, your flexibility to change that is low because it is always combined with a high hurdle that has to do with the fact that you have to demonstrate bio equivalence. You have to demonstrate comparability.

This creates the dilemma that biotech firms' excellence in applying their innovation capabilities does not necessarily entail a high level of innovations that are marketable and commercially successful (Gittelman & Kogut, 2003). In this environment, continuous customer proximity and interaction become crucial success factors. The year 1997 marked a key date in the company's recent history when Sartorius developed a new group-wide strategy focusing on growth and innovation. More specifically, Sartorius strategically aimed at expanding its technology and product portfolio by: (1) using their sales subsidiaries and local commercial agencies more efficiently, and (2) developing the business through organic sales increases and acquisitions.

In the following years, the firm acquired various companies to strengthen their strategic portfolio and reorganize the company to its present structure. At the same time, the innovation rate picked up and the company has managed to increase the scope of its innovation in terms of commercial success:

> I would think that, historically, Sartorius has always been quite innovative, but until the late 1980s it was a company that was innovative in the technology sense but not very good in terms of commercialization. I think at the moment, during the last five to six years, if I would compare our innovation rate with the one of our competitors, I would say that we are in a leading position (Senior Vice-President BP).

3.3 Business Strategy

Sartorius pursues a business strategy based on value innovation (Kim & Mauborgne, 1999, 2004). A key characteristic has been the company's transformation to a total solution provider that included a recombination of existing technologies, a complementary extension of the business line and continuous creation of new customer demand. The following sections will examine elements of Sartorius' business strategy.

3.4 Mission Statement

Sartorius has established a mission that explicitly states its focus on innovation and technology orientation. Likewise, the company has systematically shifted its overall competence from product orientation to total solution management, enabling customers to implement complex processes both in a laboratory and a production environment. In doing so, the company aligns its strategic activities towards constantly creating customer value:

> We are striving to systematically expand our position as an innovative, customer-oriented technology group. Our objective is to create lasting value for our customers and shareholders and to translate our successful growth strategy into high profitability (Sartorius mission statement).

The stated mission also reflects a corporate philosophy that views innovation as an integral part of the whole company. As a result, the company does not maintain a separate department for innovation, but rather integrates innovation as an overarching concept into the everyday processes.

3.5 Core Competencies

The company possesses two main core competencies in the biotechnology division. The first relates to the field of fermentation technology. Sartorius developed this core competence through a strategic acquisition of the market leader in fermentation, B. Brown Biotech International, in 2000. Fermentation technology reflects an upstream process and consists of the manufacturing of a protein or a biotech drug.

The second core competence lies in the field of micro-filtration. Sartorius recently extended this competence by developing additional technologies and systematically consolidating all filtration technologies into their new business unit, PT. Purification encompasses the downstream part in that it deals with isolating specific components out of a complex product. This reorganization completed the company's transformation to a total solution provider from a structural perspective, since PT served as the final segment in the whole solution package that Sartorius offers to its customers. This segment entails the highest customer value and the strongest future growth potential due to a high ratio of non-commercialized innovations.

It is evident from the above discussion that Sartorius has developed its competencies, to a large extent, through both strategic acquisitions of other companies as well as strategic recruitment of human resources that are able to provide key process know-how:

> [I]f you want to be a total solution provider, product know-how is not enough. You need process know-how, and when entering a new field you have to acquire know-how and that is mostly done either by buying a company or by buying in people that come from that field (Senior Vice-President BP).

In order to evaluate the development and extension of Sartorius' core competencies towards total solution management, the following section takes a closer look at the company's innovation strategy.

3.6 Innovation Strategy

Sartorius maintains a strategic view towards innovation. This differentiates it from a more operationally-based perspective in terms of the perception and scope of innovation:

> I differentiate between operational excellence and strategy in a way that I say "Operational excellence is to run the same race faster" and "Strategy is to run another race". And to me, innovation is more strategy, to do other things, to do new technologies, to do things differently than others do …. I would include the complete scope of activity of a company (Senior Vice-President BP).

Accordingly, satisfying existing customer needs alone does not lead to innovation. Rather, valuable innovation is thought to result from the anticipation of customers' potential needs based on the technology portfolio the company possesses or would be able to provide. This view is reflected in Sartorius' innovation strategy, namely to focus on the potential future needs of its targeted customer group.

This understanding links to current thinking of Kim and Mauborgne (1999, 2004) who illustrate the superiority of organizations pursuing what they call a value innovation strategy. The creation of new demand lies at the centre of this concept and entails not only the fundamental extension of customer value in existing markets, but also the development of potential customer needs into new markets, hence creating new customers. A key element of Sartorius' innovation strategy has been the recombination of old technologies that enabled the firm to develop innovative applications:

> We acquired BBI Fermentation Technology that was certainly a rather old technology and could not have been considered an innovative technology. That means, ... through an older technology we came to innovation because we could combine, as the only company ..., fermentation technology with membrane filtration technology. Both technologies in themselves are rather old technologies, but in the combination it is really innovative because there is nobody else who can offer that (Senior Vice-President BP).

3.7 Resource Availability

Although Sartorius disposes of only 10–20 percent of the R&D budget of its competitors, the company manages to maintain an innovation rate that exceeds its competitors. As a result, it appears as if R&D funding exerts a minor influence on the innovation rate of a company. In this respect, the Head of PT stated that

> [i]f you look at very big pharmaceutical companies ..., they are now spinning out small R&D areas that they almost treat as start-up companies because they found out that the innovation potential in terms of more ideas, more flexibility and more risk acceptance is higher in these small structures.

This notion suggests that not just available financial resources, but also the organizational structure and the innovation culture play a major role in ensuring that a firm is able to innovate. This is reflected in Sartorius' organizational structure as perceived by the Senior Vice-President of BP:

> [T]here is a certain kind of chaos and I also think that chaos at a certain degree is a good nutrient for innovation and creativity.

However, both respondents stress that resources do play a vital role in terms of transforming innovations into commercially successful products or solutions. In fact, existing organizational resources naturally constrain the amount of innovative ideas, concepts and products that can be introduced to the market and converted into commercial success. Research suggests that in addition to the necessary initial funding of innovation processes, biotechnology companies tend to ensure a steady flow of resources by maintaining strategic collaborations and alliances (Coriat, Orsi & Weinstein, 2003).

3.8   Collaboration with External Partners and Absorptive Capacity

Absorptive capacity is defined by

> the ability of a firm to recognize the value of new, external information, assimilate it, and apply it to commercial ends (Cohen and Levinthal, 1990, p. 128).

A strong body of research highlights the role of formal collaborative ties with external partners in terms of increasing the innovation output of biotechnology firms (Baum *et al.*, 2000; Powell *et al.*, 1996). In the case of Sartorius, the collaboration with external partners is of central importance for the development of innovations. In particular, the firm, in line with other biotech companies (Gittelman & Kogut, 2003), maintains a close relationship with scientific knowledge sources:

> We collaborate with universities and institutes as much as we can ....
> We absolutely believe that if we can combine different people,

know-how and disciplines, then our innovation rate goes up (Senior Vice-President BP).

The scope of external collaborations at Sartorius encompasses a wide array of activities. For example, the company focuses on close cooperation with centers of excellence in the respective scientific fields, actively contributes to the creation of new centers of excellence, finances PhD theses, organizes seminars and workshops, hosts conferences and congresses, maintains a dialogue with scientists on a regular basis and assigns contract work to certain universities. Generally speaking, a main focus lies on research institutions that are primarily driven by applied science, such as the Max-Planck Institute, Fraunhofer Institute or the Massachusetts Institute of Technology. Collaborations with other institutions also exist. The personality of the respective external partners plays an important role for Sartorius:

> [I]t is not only the name of the university, it is also the partners we are dealing with. Are we dealing with a university where the professor is open-minded or is it a person that is very academically driven? In selecting the partner, the personalities that we see on the other side have an influence (Senior Vice-President BP).

Furthermore, one respondent highlighted the need to establish several collaborations that are sustained on a parallel basis. In doing so, it is possible to recombine and integrate results and findings from different projects, thereby enhancing the potential for developing innovative solutions. This approach highlights the need to implement an appropriate formal business structure that is conducive to simultaneous innovation processes (Burgelman & Maidique, 1988).

3.9 Complementary Assets

Research suggests that firms can increase organizational learning and innovation by leveraging on their knowledge base (Cohen & Levinthal, 1990; Zahra & George, 2002). In our earlier discussion, we discussed how Sartorius develops its innovation strategy through strategic acquisitions and hiring people with expert knowledge. Similar to the collaboration with external partners, these assets enable the company to enhance its innovation capabilities:

> I get or buy in more know-how and by using that I hopefully come to innovation. But innovation is something you cannot buy in (Senior Vice-President BP).

There are other factors that facilitate innovation, mainly a corporate culture that supports innovation. The culture at Sartorius can be characterized along several dimensions. First, it entails a risk-tolerant environment. The Head of PT emphasized that

> people who are supposed to bring innovation must have the possibility to fail, and failure cannot be sanctioned in the company because if you sanction failure ..., I think, this is minimizing the drive for innovation.

Additional cultural features at Sartorius include freedom of operation and knowledge-sharing behavior that consists of an open disclosure of information. Moreover, workforce stability is crucial in order to retain, develop and continuously benefit from existing company-specific knowledge:

> [Y]ou also have to offer an environment that prevents people from leaving the company and spreading the ideas, which you cannot inhibit by any means. If people want to go, they go. I mean, even when they want to keep their knowledge, they simply cannot avoid spreading that knowledge in their new position. On the other hand, ... the fluctuation rate at Sartorius is very low (Head PT).

Finally, as will be discussed in more detail in the next section, the internal combination and diffusion of knowledge play a central role for innovation.

3.10 Innovation Capability

Innovation capability is of critical importance to the long-term success of the organization. However, it is not clearly understood what constitutes innovation capability in an organization. Lawson and Samson (2001) define innovation capability as "... the ability to continuously transform knowledge and ideas into new products, processes and systems for the benefit of the firm and its stakeholders."

3.11 Knowledge Management

A central innovation capability can be identified in Sartorius' efforts to foster internal diffusion and external exchange of knowledge. To enhance this transfer, the company opened a corporate university called Sartorius College in the year 2001. The college encompasses 2,100 m^2 of space, accommodating up to 400 people. The concept for this college goes back to the initiative of Sartorius' former CEO who emphasized the need for constantly sharing knowledge in order to reap knowledge benefits in terms of innovation and continuous improvement. More specifically, Sartorius views the college as a corporate model for applied knowledge management:

> We have to give credit ... to our former CEO ..., who was the father of creating this college. For him, college is living knowledge management, and he said, "We need to have a place, a college, where universities, customers and we meet just for increasing knowledge. Knowledge is not power, but sharing knowledge is power!" (Senior Vice-President BP).

The college boasts several places for social interaction like a bistro and a restaurant. This intends to further increase the exchange of knowledge between employees. Also, it intends not only to foster internal knowledge-sharing, but also to assimilate external knowledge and impart it to specific in-house groups. For example, Sartorius organizes an annual BioTech Forum, a conference on new trends in biotechnology that attracts both scientists as well as customers from different countries. Overall, the knowledge management strategy of Sartorius highlights that knowledge becomes more valuable in terms of innovative output the more frequently it is exchanged and recombined. In fact, empirical evidence suggests that

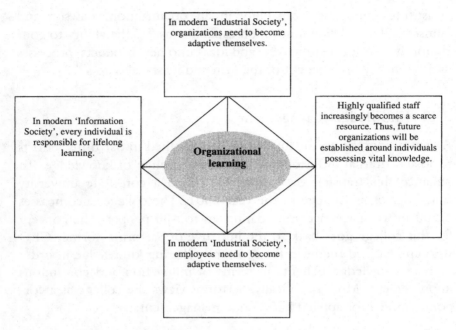

Fig. 3.1 The principles of organizational learning at Sartorius AG.

Source: Adapted from Sartorius AG.

the recombination of existing and creation of new knowledge through exchange networks can substantially leverage innovation (Tsai & Ghoshal, 1998).

The construction of the Sartorius College can be viewed as a systematic investment in the company's knowledge base. Sartorius' idea of physically gathering employees in order to exchange and integrate individual knowledge and thereby foster organizational learning is based on four central assumptions as visualized in Fig. 3.1.

3.12 Sustainable Development

Sartorius' corporate strategy reflects a responsibility towards the environment. The company has been certified in accordance with the DIN EN ISO 14001 standard. In many areas, Sartorius has developed exemplary solutions that protect both environment and resources.

Furthermore, the company demonstrates an environmental consciousness in terms of building its product portfolio:

> If you talk about disposable single-use equipment, then it always makes sense to look at the overall picture because using a piece of technology once and throwing it away after that and incinerating it can be even better, also from an environmental perspective, than to use it over and over again, because you use such an awful lot of water and other utilities to clean it and to prove afterwards that it is clean, that the balance is in favor of disposables (Head PT).

Sartorius College also offers courses on practical environmental protection and relays information about the environmental audit based on the European Eco-Management and Audit Scheme and the certification process according to the DIN EN ISO 14001 standard.

There seems to be a strong external influence on the application of practices related to sustainable development at Sartorius. More specifically, legal requirements and monetary aspects affect Sartorius' environmental consciousness:

> We are overfulfilling legal aspects because you have to anticipate the development in the future. In two or three years from now, emissions have to be further down and if you don't have that, then you are in trouble and you have to pay for that (Head PT).

However, from a customer perspective, Sartorius does not yet seem to benefit from introducing environment-friendly products into the market. Both respondents noted that customers are still mainly unwilling to pay a premium for green products. At the same time, the knowledge the company has gained from its experience in developing environmentally conscious solutions has provided the basis for building up the new business area, Environmental Technology, which is now set up under the biotechnology division. The main focus in this area lies in the development of an innovative generation of membranes for fuel cell technology.

The development of this business area provides a good example of how the company recombines existing technologies and product solutions from its biotechnology and mechatronics divisions in order to develop innovative applications. The strong linkages between the various business areas ensure cross-fertilization and continuous development of existing expertise. It becomes evident that Sartorius has made systematic use of its experience in developing environment-friendly products and processes, both through external initiatives in terms of legal requirements as well as internal recombination of existing knowledge, to facilitate the development of a new business area.

By adding environmental innovations to its total solution approach, Sartorius acknowledges the growing importance of green product development for increased operational excellence and long-term customer acceptability. This not only helps to maintain corporate competitive advantage, but also ensures sustainable development for society (Porter & van der Linde, 1995; Noori & Chen, 2003).

3.13 e-Commerce

e-Commerce and the use of the internet at Sartorius are of specific value in terms of both e-Procurement and as a potentially new distribution channel. In addition, international work groups at Sartorius interchange data using the company's intranet. However, e-Commerce cannot be a source for innovation in itself, but rather has to be viewed as an enabling tool that can facilitate the development of innovative solutions:

> If you ask "Does e-Commerce help innovation?", the answer is probably yes. Due to e-Commerce and due to the internet, you have more access to more information and so you can get more information in a shorter time and that helps you to innovate .... I think e-Commerce is a good way to increase the know-how, the information level in a company, that then potentially can lead to more innovation (Senior Vice-President BP).

This statement highlights the view that although the internet increases the availability of information, it is difficult for a company to differentiate itself through the use of this information. Rather, it is important to possess a firm-specific background of skills, experience and knowledge. This background has the potential for a sustained competitive advantage. As described earlier, the Sartorius College intends to foster systematic knowledge transfer within and beyond the company and thus aims at building up this knowledge base. This finding is supported by a growing strand of literature that views e-Commerce and the internet as an enabling technology (e.g., Porter, 2001). While the internet *per se* will rarely result in a competitive advantage, it provides companies with better opportunities for distinctive strategic positioning than did previous generations of information technology.

3.14 New Product Development

As noted earlier, transforming innovative applications into marketable products and solutions is a crucial challenge in the biotechnology sector that is especially prone to a "disconnect between scientific knowledge and important technologies" (Gittelman & Kogut, 2003, p. 380). Sartorius has addressed this risk by building a specific strength in integrating its customers into the development process. This ensures that innovative solutions are convertible into commercial success. The Senior Vice-President of BP noted that it is important

> ... working with the customer and incorporating the customer into the development, because having a powerful R&D is one thing and it is certainly a very important success factor, but without the feedback from the customer and the feedback from the processes and the applications, it is not useful in any sense because they are then developing solutions for problems that don't exist.

Empirical research corroborates these findings by showing that customer orientation in innovation projects has a positive influence on new product development success and that the impact increases with the degree of product innovativeness (Salomo *et al.*, 2003). Sartorius' strategy of being a total solution provider entails the need to focus on optimizing customer processes. The development of new products thus results from specific process-related requirements. New product development at Sartorius is therefore very much driven by the total outcome in terms of the innovative technologies that are used by customers in a combined way:

> [T]he best solution for a customer might not be to have the best products, but to have the best combination of different technologies, which doesn't mean to have 100 percent here and 100 percent there, but the total outcome must be better. That means the product development has to consider whether it makes sense to develop a product to a 100 percent performance or whether it is not better to have an 80 percent performance and another product that also performs to an 80 percent (Senior Vice-President BP).

Finally, Sartorius fosters new product development through a recombination of available technologies in order to offer innovative solutions. This ability has resulted from the firms' strategic focus on innovation through anticipation of its customers' potential future needs. Following the aforementioned example of recombining the two old technologies of fermentation and membrane filtration, the Senior Vice-President of BP noted:

> We anticipated, we were driven by the belief that this is a potential need, so we put two old technologies together, now finding out that this is exactly what the customer is now asking for because he has been optimizing fermentation technology to a 120 percent and has not addressed this benefit in the downstream process.

3.15 Organizational Performance

Sartorius closely links innovation output to market success. Accordingly, innovative solutions that are developed by its employees establish a fit with existing or potentially arising customer needs:

> [I]nnovation for a company like Sartorius only makes sense if it leads to commercial success. And you will only have commercial success with innovations if they really address customers' potential needs (Senior Vice-President BP).

Several measures of innovation do not necessarily reflect organizational performance. For example, the number of patents might demonstrate the innovative activity of a company, but this measure does not entail whether a respective patent makes sense in terms of customer needs and thus whether it leads to commercial success. Therefore, Sartorius places a specific focus on the feedback and evaluation the company receives from its customers. As the Head of PT noted,

> The measurement that I think is relevant is something we have seen now — the way customers and the environment acknowledge us and how they perceive our new technologies.

However, the main criterion used for measuring continuous innovation at Sartorius is the percentage of sales of new products, thereby encompassing the market success of a specific innovative solution. While this appears to be an adequate measure from an organizational viewpoint, it does not necessarily reflect innovation with regards to the customer perspective:

> In terms of the customer, that might be completely different because they might have had this product already available for ten years from a competitor and now we are coming out with a similar product (Senior Vice-President BP).

Sartorius uses common measures to evaluate its organizational performance like "Earnings before Interest and Tax" and "Return on Capital Employed". The huge investments that Sartorius have made during the last years, for example concerning a new manufacturing plant and the Sartorius College, have decreased the disclosed return on capital employed.

These investments have to be viewed in line with the company's conversion from a product-oriented company to a total solution provider:

> We are in this phase where you will only see the results of our total solutions approach in terms of figures in five or ten years .... You cannot convert a company from a product to a solution provider in two or three years. You need to change a complete culture and that takes at least ten years, at least, so that it is unchangeable again (Senior Vice-President BP).

Accordingly, Sartorius embraces a longer-term perspective towards its performance measurement and monitoring. At the same time, this longer timeframe seems to be a constituent of highly innovative organizations. As the Senior Vice-President of BP stated,

> companies that are highly innovative are usually not those companies that have good economic figures. Most of the companies that turned out to be very innovative had very bad economic figures at the beginning and later on very good ones .... Innovation is something that secures the future existence of a company. All the economic figures are measuring something that happened in the past.

Finally, the company considers aspects of social responsibility in its performance evaluation with regards to a broad range of stakeholders:

> [W]e are not a start-up company and we are responsible for 1,700 people in Goettingen. This is also an evaluation factor whether, at the end of the day, we have been able to feed and pay all these people and are still profitable so that other stakeholders are also pleased by the performance (Head PT).

3.16 Human Resource Management

Sartorius' focus on developing innovative solutions that anticipate potential customer needs requires a highly qualified and innovative workforce. It is important that employees are close to the market in order to understand customers' processes and their requirements. This entails the need to continuously train and develop the workforce, readjust job profiles and enable employees to rotate between different job scopes and processes to facilitate multi-skilling. The company has addressed this requirement by actively promoting a knowledge-sharing culture. It is based on the belief that innovative activity involves the need to facilitate knowledge exchange between organizational members. Since knowledge primarily resides within the individual (Grant, 1996), employees have to be motivated to share individual knowledge. In this regard, the inauguration of the Sartorius College has been a major facilitator. In addition, the company retains its experienced and innovative employees by providing not only extrinsic but, more importantly, intrinsic motivation. In particular, the flexible job design prevalent at Sartorius enables its employees to accompany innovation projects from the development of the innovative ideas until their commercialization. This is an important motivator to retain the company's innovation champions in the long-run (Cappelli, 2000), thereby ensuring a reliable staff base for initiating new innovations.

3.17 Customer Orientation

A systematic integration of customers into the process of developing innovative solutions is a critical means of ensuring customer satisfaction and increasing organizational intelligence, as it enables a company to better tailor its technologies to customers' needs (Burgelman & Maidique, 1988). Sartorius addresses this need by establishing a continuous dialogue with its customers, by inviting customers to seminars and conferences and organizing regular customer focus group meetings. However, a strong focus on the current customer base entails the risk of not considering the potential needs of other customer groups that have not been addressed yet. This might restrict the scope for innovative activity. Also, dynamics in the customer base involve a high effort of regaining specific knowledge on varying customers. Additionally, this includes the risk that technological solutions become too specific and thus lack generalizability towards other customers.

3.18 Stakeholder Management

The company's transformation into a total solution provider, together with the required investments, entails negative performance consequences in the short-term. As the change and the required realignment of the corporate culture is an ongoing process, the entire benefits from the conversion will only unfold in the long-run. In the meantime, the company faces the challenge to motivate and satisfy both its shareholders and other stakeholders. The latter include Sartorius' employees who need to be aware and supportive of the change process. Here, a clear communication of the business strategy and active participation of various stakeholder groups in the change process are essential for success.

3.19 Conclusion

This case documents innovation capabilities prevalent at Germany-based Sartorius AG. More specifically, the analysis demonstrates that business strategy is a major determinant for leveraging the company's innovation capabilities. Sartorius pursues a business strategy of value innovation and focuses on the anticipation of existing and potentially new customers' future needs, based on the technology portfolio the company possesses or would be able to provide. At the same time, Sartorius' strategic transformation from a traditional product-focused company to a total solution provider exerts a major impact on the way the company is managed and structured. Three major innovation capabilities can be derived from the case analysis. First, Sartorius actively promotes constant intra-company knowledge diffusion as well as collaborative ties and interchange with external partners. This is achieved through the company's corporate university that institutionalizes knowledge transfer within and beyond the company's boundaries, and also through a deeply embedded knowledge-sharing culture that rewards exchange. Both elements provide exchange networks that ultimately leverage innovation.

A second, related innovation capability is the recombination of existing expertise and technologies to develop innovative solutions. For example, Sartorius has systematically exploited the experience it has gained in developing required environment-friendly solutions to facilitate the development of a new business area. Also, the company recombined two existing technologies and extended their application to problems that could not be tackled by either of the two technologies alone. In doing so, Sartorius not only increased customer value, but essentially created new demand. A third innovation capability lies in the company's effort to systematically integrate its customers into the product development process. This close interaction ensures that the innovative output is linked to a measurable value that derives from providing a fit with customer needs and thus leads to commercial success.

The main implication of our case study for managers and academics is that companies need to align their structure in a way that enables them to maintain and leverage innovative output, customer value and commercial success, even though this process might entail negative performance outcomes in the short-term. The case also presents important implications with regards to how innovation is measured. Clearly, the number of patents does not adequately reflect the innovation rate of a company. Rather, innovation needs to be assessed in terms of its market success. Innovations can only be regarded as such if they add value in terms of how existing and potential customer needs are addressed. Finally, the link between innovative output and commercial success highlights the importance of organizational resources that can be tapped to bring innovative ideas and solutions to the market. In the case of Sartorius, and certainly other organizations as well, this marks a critical bottleneck as the resources to commercialize innovative technologies are limited.

Review Questions

(1) Discuss how Sartorius formulates its innovation strategy based on the value innovation concept.

(2) Explain the role of Sartorius' corporate university. Do you think Sartorius has been able to gain an appropriate return on their investment? Why?

Chapter 4

Managing Strategic Change Through Mainstream and Newstream Innovation at Eurocopter, France

Milé Terziovski and B. Sebastian Reiche

4.1 Introduction

Recently, there has been an increased research interest in a dynamic capabilities perspective of innovation (Teece & Pisano, 1994; Galunic & Rodan, 1998; Eisenhardt & Martin, 2000). Building on the resource-based view of the firm (Wernerfelt, 1984; Prahalad & Hamel, 1990; Barney, 1991), these researchers argue that competitive advantage is a function of a firm's dynamic capabilities that enable a continuous exploitation and recombination of idiosyncratic resources to create new products and processes. Along these lines, scholars have high-lighted the need to not only efficiently manage a firm's mainstream processes, but also develop complementary newstream activities that enhance innovation (Kanter, 1989; Lawson & Samson, 2001). A key focus thus lies on the identification of those innovation capabilities that are able to leverage new value creation.

The aerospace industry serves as an adequate field to study innovation capabilities for two main reasons. First, with ever more players entering the market, the industry has undergone a substantial increase in competitive scope over the past decades (Jurkus, 1990; Haque & James-Moore, 2005). Second, the military aircraft market is highly dependent upon national defence budgets (Anand & Singh, 1997). A volatility of these budgets in response to changing political conditions results in the industry facing constant pressures to retain existing and acquire new customers.

The sector has been subject to a variety of research, both quantitative (Greer & Liao, 1986; Walls & Quigley, 1999) and qualitative (Prybutok & Ramasesh, 2005) in nature. Adopting a qualitative single case study approach (Eisenhardt, 1989; Yin, 2003), the present study examines sources of innovation capabilities in this industry segment, focusing particularly on France-based Eurocopter. Specifically, we analyze the role of process innovation, customer orientation and strategic networking as key mechanisms to achieve an integration of mainstream and newstream activities that thereby enhance the firm's innovation capabilities.

4.2 Case Study Interview

The case study research was conducted by Associate Professor Milé Terziovski through an in-depth interview with a senior manager from the Department of Competitiveness at Eurocopter in La Courneuve, France. The interview was conducted in French and immediately translated into English with the help of a professional interpreter who attended the conversation. The interview was complemented with other information such as various company brochures, the company website and annual reports in order to ensure data triangulation (Miles & Huberman, 1994; Yin, 2003).

The text is divided into four sections. First, the case company is characterized in terms of its corporate background, core competencies, approach to innovation and resource availability. Second, we present a conceptual model that serves as a framework to analyze our data. Specifically, we view process innovation as the result of an integration of mainstream and newstream activities. We then focus on the key elements of innovation capability at Eurocopter. We investigate the firm's two-stage innovation strategy in terms of: (1) a realignment of its core business processes of blade repair, and (2) establishing systematic customer involvement as a means to create increased customer value. This also entails an analysis of the firm's strategic network of external partners as an important source of innovation. The text concludes by highlighting organizational implications and drawing together the main findings.

4.3 Company Background

Eurocopter is the world's leading helicopter manufacturer, holding 60 percent of the global civil and public market. The company was formed through a merger of the helicopter divisions of German DaimlerChrysler Aerospace AG (Dasa) and the French aerospace company, Aerospatiale Matra. Eurocopter is a wholly-owned subsidiary of EADS (European Aeronautic Defence and Space Company), which, by employing more than 100,000 employees at more than 70 production sites in Europe, is the largest aerospace company in Europe and the second largest in the world. In 2002, Eurocopter generated sales of 2.5 billion euros, with 58 percent stemming from the civil market and 42 percent from the military sector. The company has a workforce size of around 11,000, distributed across its three plants and facilities in Germany, two in France and 15 additional subsidiaries and participations worldwide.

The group's headquarters is based in Marignane, France. Eurocopter develops, manufactures and markets a wide range of helicopter models that, in total, cover about 90 percent of the global helicopter market requirements. The site in La Courneuve, the unit of our analysis, has been established as a competence center for the design, manufacturing and maintenance of composite rotor blades, propellers, turbine blades and other products with blade repairs, which accounts for 25 percent of total sales. The site employs around 750 employees and has over 1,500 customers in more than 120 countries. While helicopters boast a fairly long lifecycle of about 40 years, blades usually need to be replaced earlier. This entails the need to ensure continuous customer support through process and product innovation over the entire product lifecycle.

4.4 Core Competencies

The site in La Courneuve possesses two main core competencies. First, the plant has developed its own technology with regards to manufacturing and repairing rotor blades and related products. The fact that the design and manufacturing stages are closely linked has prompted the company to set up an integrated process that incorporates computer-aided design, large-scale production of machining and core materials, and the application of leading-edge technology to manufacture complex component parts and extensive test facilities.

As we will show below, the reengineering of these core business processes has enabled the company to increase its innovation capabilities. The successful realignment has been publicly acknowledged by awarding the company with the Quality Aerospace Award in 1999. Secondly, our respondent highlighted the importance of the firm's human resources in terms of their idiosyncratic knowledge and commitment:

> It is very important for us that our employees have an in-depth knowledge of the product and that they also like and identify themselves with it.

This knowledge serves as an important source of creating increased customer value. Indeed, research suggests that firms can increase their innovative output by leveraging existing knowledge that resides in its employees (Cohen & Levinthal, 1990; Grant, 1996).

4.5 Innovation Approach

The company does not locate innovation in a specific department, but rather sees it as a process that is prevalent in everyday routines:

> There is no specific position for innovation in the company. It cuts across all functions.

As a result, innovation becomes a state of mind that guides both employees and corporate decision-making. Specifically, innovation as perceived by Eurocopter encompasses various elements. From a process perspective, innovations occur in terms of a reorganization of work processes to manufacture component parts and blades. From a product perspective, innovation is perceived as the outcome of application research that ultimately leads to innovative product or parts development. The Eurocopter site in La Courneuve makes extensive use of cross-functional teams in order to stimulate innovation. In this regard, staff from the customer support side is linked up with employees from the product design and quality departments.

4.6 Resource Availability

Eurocopter maintains a wide array of research and development projects that are resourced with approximately 10 percent of annual sales. A key focus of R&D is associated with noise abatement. As the respondent noted,

> Our challenge is the noise factor. We need to reduce the noise at least by 10 percent.

Other R&D projects entail work on full all-weather capability of helicopters and the potential for optimizing in-flight comfort and the safety of pilots and passengers. Finally, the company faces external pressures with regards to sustainable development. Specifically, European legislation requires the company to focus on waste minimization and the use of environmentally friendly products. This results in certain firm resources being allocated towards meeting environmental requirements. Next, we will briefly describe a conceptual model of innovation that guides the following in-depth analysis of the case data.

4.7 Mainstream and Newstream Capabilities

Fig. 4.1 illustrates the integrated model of innovation. Although organizational competitiveness has long been assessed in terms of how well companies manage their corporate mainstream variables such as efficiency, quality, customer responsiveness and speed, these activities only represent a necessary but not sufficient condition to ensure a sustained competitive advantage in today's business world (Kanter, 1989). Accordingly, scholars and practitioners alike have extended their perspective in order to identify capabilities that enable companies to continuously innovate and actively shape their existing and future markets. Specifically, it has been suggested that an organization needs to complement its mainstream capabilities with newstream capabilities in order to be able to initiate innovative business streams.

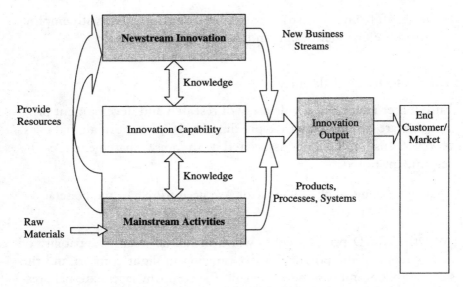

Fig. 4.1 An integrated model of innovation.

Source: Adapted from Lawson and Samson (2001).

While mainstream activities provide the necessary stability to maintain organizational functioning through process innovation, the newstream activities introduce a dynamic context that requires continuous new product development (NPD), as well as knowledge creation, application and recombination. In doing so, the organization is a constantly moving target to competitors (Kiernan, 1996). Building on these ideas, Lawson and Samson (2001) argue that mainstream and newstream activities need to be managed in an integrative manner to achieve innovation. This can be achieved by creating innovation capabilities which, in turn, are able to combine key resources and capabilities to initiate innovation (Fuchs *et al.*, 2000). In this respect, a firm's mainstream operational activities provide the necessary tangible and intangible resources to develop future business streams.

Supporting this notion, Terziovski (2002) shows that a combination of continuous incremental improvement and radical innovation can result in increased performance outcomes — newstream innovation. We now turn our attention to the newstream innovation at Eurocopter. We shall define "newstream" as the aggregate of a firm's resources that are geared towards identifying and creating new value for customers. First, we examine how Eurocopter realigned its core business process of blade repair to achieve continuous improvement in its mainstream activities. Second, we analyze how the firm established systematic customer integration in order to develop new business ideas and provide increased customer value. Last, we illustrate the important role of strategic alliances as a source of newstream innovation at Eurocopter.

4.8 Stage 1: Innovation of the Blade Repair Process

In the 1990s, Eurocopter embarked on a continuous innovation project to improve its mainstream activities with the ultimate aim of ensuring high quality, increasing customer satisfaction and improving firm performance. In doing so, the firm changed from a sequential vertical approach to horizontally aligned cross-functional processes. The blade repair process serves as a good example to illustrate this change. The initial repair process was very much isolated from the customer once a blade was handed in for repair, with the blade passing sequentially through the warehouse, technical assessment, repair/ maintenance and customer support until it was delivered back to the customer (see Fig. 4.2). In this configuration, technical assessment represented a critical bottleneck because the product-related defects and customer preferences in terms of blade maintenance had to be identified in this department. This led to a substantial amount of waiting time in the blade repair process, which led to an increased blade repair cycle time (BRCT). A change to a horizontal structure that involves customer feedback and concurrent processes across functions enabled the firm to reduce cycle times substantially. Our respondent emphasized that

> through continuous innovation of processes, we reduced the cycle of technical assessment of blades from initially 26 days to 14 days.

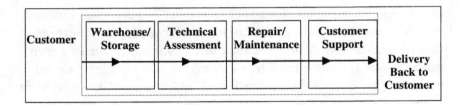

Fig. 4.2 Vertical approach to blade repair.

Specifically, the new configuration allows problems and customer preferences to be clarified through the customer support department. After this initial step, the blade moves to the technical assessment which, in turn, only focuses on an assessment of the necessary repair/ maintenance work to be conducted. Based on the initial clarification of customer preferences, technical assessment is able to develop a precise forecast in terms of price and duration, which is then fed back to the customer for appraisal. If the customer agrees with the conditions, the blade is sent to the repair/maintenance unit.

The realignment of processes also entails a closer involvement of customers. The systematic use of customer feedback and the resulting integration of customers into the product development process provide the foundation for initiating newstream business activities at Eurocopter. An important element of the company's customer orientation can be derived from the fact that the company is more concerned about product feedback from customers than benchmarking itself with competitors. The respondent emphasized:

> We don't just benchmark our products with our competitors because we feel that we can always catch up with them. What is more important is to understand our customers and their future needs.

Research supports these findings by showing that customer orientation in innovation projects has a positive influence on new product development success and that the impact increases with the degree of product innovativeness (Salomo *et al.*, 2003; Terziovski, 2001). To achieve customer involvement, the site in La Courneuve makes extensive use of customer surveys and customer focus group meetings. However, Eurocopter went beyond involving customers throughout the blade repair process. The firm created a customer philosophy that concentrates on creating and satisfying future customer needs and thus actively initiates innovation newstream:

> Our strategic goal is to realize and satisfy new customers, open new markets, ... (Formula 12, Strategic Goals Statement).

This perspective ties in with recent literature on value innovation, suggesting that those firms that maintain their existing customer base develop future business and thus, create new customers that remain loyal to the organization (Kim & Mauborgne, 1999, 2004).

The combination of reconfiguring the core processes and decreasing manufacturing cycle times on the one side and providing new services to customers on the other were decisive for the company in winning the Quality Aerospace Award in 1999.

4.9   The Role of Strategic Alliances

The use of external partners and alliances is an important source of newstream innovation as it allows the firm to tap into additional innovation capabilities external to the organization and recombine these with its own capabilities (von Hippel, 1988; Gulati & Garino, 2000). Benefits of knowledge exchange through external networks appear to be particularly beneficial to innovation outcomes if network members are not direct competitors (Bouty, 2000). At Eurocopter, the use of strategic networks and partners occurs along three main dimensions.

First, the company is involved in various joint manufacturing projects with other companies. For example, Eurocopter manufactures the twin-engine helicopter BK 117 in cooperation with the Japanese manufacturer, Kawasaki. Kawasaki supplies the cabin and electrical systems, while Eurocopter produces the tail boom, the dynamic system and the engine equipment. The collaboration is considered to have resulted in a helicopter model featuring superior technology, reliability and flight performance.

Second, Eurocopter maintains joint R&D facilities and collaboration with other industrial partners in the aerospace industry. In addition, the firm is in close contact with two aerospace institutes, the German Aerospace Center (DLR) and the French Aeronautics and Space Research Center (ONERA), respectively. As a result, the company sustains several research collaborations on a parallel basis. This approach emphasizes the need to implement an appropriate formal business structure that is conducive to simultaneous innovation processes (Burgelman & Maidique, 1988).

Third, Eurocopter places great importance on close collaboration with its suppliers in the design process of blades. For example, joint discussions with suppliers help to clarify which composite materials are to be used and which part of the development process can be outsourced to suppliers. This ensures that the final product provides leading-edge quality to the customer:

> In the past, we only focused on our two major suppliers, but today we want to be on good terms with all players. We want to create the highest degree of synergies possible. When we develop new products, we consult with all our suppliers to decide who will develop a specific material and what the design would look like. We then decide whether and to what extent we outsource a specific process.

In doing so, the company aims to create a win-win situation where close collaboration benefits both Eurocopter and its suppliers by achieving synergies, continuous feedback and long-term cooperation. The importance of international and interfirm networks at Eurocopter is a result of the company's history. Indeed, as mentioned earlier, the company was created through a French-German merger, with the parent company EADS itself emerging from an integration of three aerospace companies from different European countries.

4.10 Interaction of Mainstream and Newstream

This case study demonstrates that both mainstream and newstream activities need to be managed interdependently in order to stimulate and sustain innovative output. While newstream capabilities clearly possess the potential to identify and create new value for customers, a company must ensure that the resulting innovations can be implemented and brought to the market. In this context, the mainstream business remains critical as it is responsible for the underlying processes and systems that maintain the company's interface with its customers and the market (Lawson & Samson, 2001). More importantly, while innovation introduces instability into the system by creating new ideas whose results are still uncertain, the mainstream capabilities ensure efficiency and effectiveness in the business processes and are the main source of resources that are allocated towards the newstream. In the case of Eurocopter, the continuous innovation of the firm's core business processes, with the resulting decrease of cycle times and increase of efficiency, provide the foundation for customer satisfaction and quality. Although these activities achieve continuous incremental improvement, they will not initiate more radical innovation. The creation of new and sustained customer value, in turn, only emanates from the leveraging of resources through systematically integrating customers and establishing a wide array of external relationships.

4.11 Human Resource Management

The success of innovating the core business processes and establishing systematic customer integration was, to a considerable extent, dependent upon contingency factors such as work environment and employee involvement. Through the use of various HR tools such as empowerment, the site initiated a far-reaching change in the mindset of its employees from a functional focus towards integration and customer orientation.

First, the site introduced employee participation and empowerment in order to create strong support for the change program. As the respondent stated,

> We needed to be very close to the employees and explain ourselves every day.

Second, there is a need to constantly train and develop staff in order to create a dynamic context for continuous product improvement:

> We need to retrain our employees and adapt their competences in order to continuously improve our products.

An important implication of nurturing this knowledge culture is the notion of knowledge exchange. In this respect, scholars highlight that employees do not only increase their knowledge base through company input in the form of training and development, but also through the systematic sharing of individual knowledge amongst colleagues. This in turn can be an important determinant of competitive advantage (Argote & Ingram, 2000). Also, research suggests that access to heterogeneous knowledge within the company is crucial for innovation (Rodan & Galunic, 2004). In this regard, Eurocopter facilitates communication, interaction and support between the various functional areas. Additionally, by leveraging individual knowledge and sharing it among the workforce, the company is able to prevent the loss of unique knowledge that occurs through employee turnover.

Third, the site facilitated the change in mindset among employees by creating team spirit and systematic job rotation:

> We suppress administrative tasks to create a motivation of team spirit. At the same time, we encourage our employees to view the big picture rather than narrowly focusing on specific work processes.

This has increased both the flexibility of human resource allocation as well as employees' extended understanding of overarching corporate processes. The latter is particularly instrumental with regards to fostering creativity and generating ideas with the ultimate aim of initiating innovative output (Lawson & Samson, 2001).

4.12 Conclusion

Drawing upon an integrated model of innovation, this chapter analyzes innovation outcomes that result from the interplay of mainstream and newstream firm capabilities at French-based Eurocopter.

Specifically, the case indicates that both mainstream and newstream activities need to be managed in an integrated fashion in order to sustain innovative output. In addition, a supporting HR infrastructure that facilitates employee empowerment, knowledge-sharing and teamwork is crucial. An important element of Eurocopter's success has been the optimization of its mainstream process innovation activities. Embarking on a large-scale process innovation project, the company was able to reengineer and realign its core business processes. The example of the blade repair process shows how the company moved from a sequential vertical structure to a horizontal approach with continuous customer interaction. This customer involvement, in turn, became part of a more far-reaching change in customer philosophy at Eurocopter.

The case demonstrates how a systematic integration of customers into the product development process provided the company with a key capability to initiate innovations by creating new and sustained customer value. This perspective highlights the shift in corporate focus from competitor-based benchmarking to radical customer orientation, as suggested in various strands of current innovation research (e.g., Kim & Mauborgne, 1999, 2004). The analysis also confirms the need to exchange knowledge and resources across organizational boundaries. Indeed, networks and alliances with key customers, suppliers, competitors and other participants help integrate complementary innovation capabilities, thereby fostering the development of new business streams. This is particularly relevant in high-technology environments such as the aerospace industry where firms will not be able to maintain competences in all potentially important technical areas.

4.13 Implications for Managers

Several implications for innovation research and practice can be derived from the analysis. First, companies need to place equal importance on both mainstream and newstream capabilities in order to initiate and sustain innovative output. In addition, as these two streams provide complementary but interdependent resources, they need to be integrated. This means that newstream capabilities without a supporting mainstream structure are unlikely to stimulate innovation. Accordingly, innovation cannot be confined to a specific functional or positional part of the company. Rather, it has to be incorporated into an organization-wide mindset that underlies all business processes. Furthermore, the case also points towards a broader perspective for locating critical innovation capabilities. This finding emphasizes the need for managers to reach beyond their immediate organizational boundaries and find additional sets of resources that can support both product and process innovation. More related research is needed to determine whether specific configurations of mainstream and newstream activities result in different innovation outputs.

Review Questions

(1) Discuss how Eurocopter managed "mainstream" and "newstream" activities interdependently. Can they exist separately?
(2) Explain why Eurocopter changed from a vertical to horizontal structure. What are the benefits that resulted from the new structure?

Chapter 5

Leveraging Innovation Capabilities at Caterpillar Underground Mining (UGM) Pty Ltd

Milé Terziovski and Ordan Andreevski

5.1 Introduction

UGM and three other specialist players dominate the global industry for the design, manufacture, sales and service of underground mining and earth-moving equipment. The structure of the industry has been shown to influence the organization's ability to innovate (e.g., Holak *et al.*, 1991). Innovation at UGM has been driven by the entrepreneurial culture of the firm, which seeks to meet or exceed the needs of customers and regulators who are trying to satisfy increasingly demanding occupational health, environmental and safety issues. As a unit of analysis, UGM qualifies as a suitable case study in innovation management because it successfully makes use of its innovation enablers and capabilities to achieve its innovation performance objectives. This case study draws on the literature on innovation management (e.g., Damanpour, 1991; Eisenhardt & Martin, 2000; Lawson & Samson, 2001) and examines the internal and external factors that influence innovation. The case study analyzes the key drivers and barriers to successful innovation. The case is based on in-depth interviews conducted by Associate Professor Milé Terziovski with senior managers and the CEO of UGM. Background information on the company was extracted from the company website, brochures and annual reports which serve as a means for data triangulation (Miles and Huberman, 1994).

5.2 Company Background

UGM is part of the global Caterpillar group. Caterpillar products and components are manufactured in 50 US facilities and in over 60 other locations, in 23 countries around the globe. In 2003, Caterpillar continued to maintain its position as a global supplier of underground mining equipment, with approximately half of all sales to customers outside of the US. Caterpillar's global locations and dealer networks are key competitive advantages that have made Caterpillar the world leader in all its businesses. UGM has a well-resourced product development department. The company benchmarks its performance to other organizations in the industry and measures innovation capability as percentage of R&D spending to sales, successful Industry Research Development grants and introduction of new features. The organization's structure is flat and horizontal (maximum of four levels from the shop floor to the MD), with cross-functional management teams and cross-functional Six Sigma project teams:

> ... the structure of the company is that there are five departments, and they're traditional as you would expect in an organization like this. We've got marketing, manufacturing, product development, admin and finance, and everything else lumped into resources. Product development is split further into engineering, resources and product maintenance.

The main products/services provided by UGM are customized design, manufacturing and assembly, sales and support of underground mining equipment. The company's key customers are mining companies and earth excavation contractors. In Australia, Caterpillar has six regional dealers. The level of support that UGM and its dealers provide to customers is best illustrated by the fact that there are over 350 field service fleet vehicles; 98 percent of all Caterpillar parts are supplied within 24 hours anywhere in Australia; 95 percent of stock parts (a stock part is any part that is requested at least twice a year) are supplied over the counter, direct from the Caterpillar dealer's own stock.

The Melbourne Distribution Center is linked online to 23 other Caterpillar distribution centers around the world. UGM is the only large machinery supplier in Australia that has its own earth-moving equipment manufacturing facility, which allows it to customize its products and services to meet individual customer needs. For example, UGM builds cabins suitable for Australian conditions so as to ensure operator comfort in the harsh underground climate. UGM uses the latest manufacturing technology. The manufacturing process employs hi-tech systems for plasma cutting, state-of-the-art paint booths and computerized welding robots.

Caterpillar's Melbourne Training Center is recognized as the best in the industry because it provides professional, leading-edge, practical training and demonstrates Caterpillar's commitment to the continual development of its employees who service the dealer network, and to the dealer staff who support an ever-widening product and customer base. The Training Center's courses provide dealership personnel with the skills and knowledge to meet the full range of customer needs. Training enables dealerships to offer consulting services to assist customers with everything from selecting appropriate equipment to providing on-going product support. Courses cover efficient, productive and safe equipment operation, professional diagnosis and service, and competency in the skills required to operate a timely and accurate parts supply system.

Backing each and every UGM machine is a vast product support network to keep them running anywhere, anytime. The Melbourne Distribution Center's mission is to help customers keep their equipment operating at the lowest possible cost. The Melbourne facility carries large spare parts inventory levels to suit customer needs. For example, UGM understands that when a customer needs a part, the one it wants is the only part that matters. The part is shipped to customers as quickly as possible, almost always within 24 hours. To do so, Caterpillar has a system called the "High Velocity Product Support", which ensures the fastest possible response to customer needs through advanced network technology, sophisticated logistics systems, superior service, quality parts and total commitment. Melbourne has online computer links with dealers and other Caterpillar distribution centers around the world.

5.3 Perception and Definition of Innovation

The organization measures its ability to continuously innovate by its sales growth and number of new products and features commercialized. Innovation is perceived as a broad and multi-dimensional process involving products, processes and strategies. For example, innovation extends to financing, whereby UGM offers innovative financing options through its Financial Products Division:

> One of the other things that we believe we lead the industry in is introducing new technology, new ideas to our industry. So there's a long list of equipment innovations in underground mining that we've introduced ahead of our competitors, and have subsequently been copied by our competitors. So to us, that's a measure that we've been innovative, ahead of our competitors.

Innovation also extends to sustainable development. UGM is dedicated to both sustaining and improving quality of life. UGM is guided by its Code of Worldwide Business Conduct to meet or exceed local environmental regulations, develop solutions to customers' environmental challenges, advocate free trade and take the lead in the business community on important issues:

> One of our key competitive advantages is that we are seen as the leader in technology and innovation and product safety, and obviously we want to keep that.

UGM is committed to generating attractive returns for its shareholders. Strategic growth initiatives involving its machine, engine and service businesses are expected to drive these returns over the next several years.

5.4 Innovation Strategy

Caterpillar's vision is to be the global leader in customer value. "Innovation" and "Innovative" appear in the organization's mission statement and have been part of the company's management culture from the start:

> Caterpillar will be the leader in providing the best value in machines, engines and support services for customers dedicated to building the world's infrastructure and developing and transporting its resources. We provide the best value to customers.

Caterpillar believes that a great company is one that produces innovative products and services, has a passion for satisfying customers, has a long history of financial strength and integrity, and has the right people who are proud, confident and who show leadership. At Caterpillar, the company is proud to be recognized worldwide as a great company, and even prouder to be continuing to earn that distinction.

Caterpillar people seek to increase shareholder value by aggressively pursuing growth and profit opportunities that leverage their engineering, manufacturing, distribution, information management and financial services expertise. Profitable growth is a key goal. Caterpillar seeks to provide its worldwide workforce with an environment that stimulates diversity, innovation, teamwork, continuous learning and improvement, and that rewards individual performance. Developing and rewarding people is part of its growth strategy. Caterpillar is dedicated to improving the quality of life while sustaining the quality of our earth. Social responsibility is encouraged:

> Caterpillar has a real three-way partnership with its customers and dealers, and we're continually striving to strengthen that. A strong dealer network, Caterpillar's longevity in Australia, an unbending desire to meet and exceed customer expectations and a unique manufacturing capability all add up to an unmatched competitive advantage which gives real benefits to our customers.

5.5 Core Competencies/Innovation Capability Development at UGM

UGM has an excellent skill base and innovation management system for all its products. This helps UGM to position itself as the global technology leader for all its products. UGM's core competence is its ability to consistently deliver superior performance, quality and value to customers across a broad range of market segments:

> It's why we say ... when you buy the iron, you get the company. It's delivering the promise, forming partnerships with customers to provide them with maximum value.

Delivering customer value is a key driver throughout UGM's operations. Continual research into customer needs and the value it places on products and services help the organization form a clear vision of industry direction. Bringing together a focus on customer needs and the increasing use of sophisticated technology, engineering, continuous improvement and R&D provides the synergy to develop and implement appropriate solutions. With such a wide representation of equipment across a variety of markets and customers, UGM is perfectly placed to benchmark across industries and customer bases. UGM and its dealers recognize that customer needs are different from market to market, and even within market segments. It is a focus on these requirements that enables it to truly "deliver the promise" to customers in a consistent way.

"Delivering the promise" has always been and will continue to remain a primary focus at UGM. That does not mean UGM is standing still — far from it. UGM has already taken the next step, moving beyond the "customer satisfaction" benchmark to "value measurement" of whole-of-life performance and costing. The new millennium will see some truly exciting developments under the Caterpillar logo. At every mine site, for every customer, UGM's major goal is to optimize productivity and lower costs. Hence, when a customer buys a CAT machine, he or she is getting a total system designed to maximize return on investment. That investment begins with the machinery. Of all equipment in Australian mines, 70 percent is Caterpillar, simply because its package provides the best solutions. In large, off-highway trucks, 91–240 tonnes, Caterpillar has a lion's share of the market.

Success for UGM and its dealers comes from working with customers both before and after sales of equipment to ensure that the essential goal of productivity at the lowest cost is met. The UGM support systems begin with total on-site mine studies to identify challenges and find best-fit solutions. Australian miners operate with one of the highest machinery utilization factors in the world, applying the most intense scrutiny to cost per tonne and are amongst the most demanding in their expectations of value for money. UGM responds to that, developing a number of programs to help customers maximize efficient operation such as Equipment Investment Analysis, Oil Analysis System and others. These in turn are supported by a range of customer support programs including Project Management, Customer Forums and Alliance Programs. This "support package" makes UGM the preferred choice for many mining operations. Caterpillar has a library of case histories which detail savings made and productivity gained through the programs.

Innovation Strategy

A key feature of UGM's innovation strategy has been its continuous investment in mainstream and newstream innovation capabilities and the sharing of knowledge across functional teams. Mainstream refers to the organization's ability to stay in business, whereas newstream refers to the organization's ability to innovate new and improved output ahead of the competition (Lawson & Samson, 2003). The organization plans and allocates resources and funding for innovation each year to meet its business strategy:

> We were careful not to develop new customers unless we had the capability to support them with products in the field, or to go out and search for customers and leave them high and dry. So in terms of innovativeness, one of the key things is being very customer-focused, making sure that we were coming up with innovations that customers wanted, rather than innovativeness for innovativeness.

Using Six Sigma, Caterpillar has achieved gains in virtually all key areas of the company. In 2003, Caterpillar expanded the reach of Six Sigma into its extended enterprise. 97 Caterpillar dealers and approximately 240 suppliers are using Six Sigma to help improve their businesses, which strengthens Caterpillar's value chain. Learning is key to Caterpillar's growth and profitability.

The global commitment to learning is evident in Caterpillar University, which was established in 2001 to develop its people and ensure success in global learning initiatives in areas such as Six Sigma, leadership, manufacturing, information technology, human resources, engineering and marketing. The global product manager for learning oversees more than AUS$100 million in enterprise spending to make learning a strategic asset for improving performance and profitability. Caterpillar has invested over AUS$2 million in a new training center to provide professional, practical training for dealers in services, parts and marketing, and for its own staff in these and allied disciplines.

5.6 Resource Availability

UGM has a ten-year plan for new product development structured on planned resources for the next ten years, based on growth and sales. Each new product has its own budget, based on its Net Present Value (NPV):

> Once we're happy with that ten-year plan, then we do each year's business plan. We work out what we've got to achieve that year, what the funding would be for that, and approve that for part of our overall business plan for the year. In addition to that, of course, are the funds required for on-going product maintenance.

The literature review has highlighted that collaboration with external partners can increase the innovation output of firms (Baum *et al.*, 2000; Powell *et al.*, 1996). In the UGM case, collaboration with external sources is limited to one R&D joint venture. Its product development managers are responsible for external collaboration. UGM makes use of complementary assets such as suppliers and customers to discuss the performance of existing and the characteristics of new products. UGM has also been involved with a Brisbane-based Cooperative Research Center of Excellence in Mining:

With a new project or a new product, part of the research phase is visiting customers, so we would usually send the two leading engineers on the project on customer visits to get their input, advice and requirements before we even start. Also, the supply team will work with key suppliers. In the case of a large underground truck, there wasn't a suitable tyre, so we worked with a tyre supplier to develop a new tyre.

The scope of UGM's collaborative effort extends to working with the CSIRO's (Commonwealth Scientific and Industrial Research Organization) former Mining Division, the Australian Metal Industries Research Association and a small technology company that is working on automation products. The literature review has found that organizations can facilitate organizational learning and innovation by leveraging their knowledge base (Cohen & Levinthal, 1990; Zahra & George, 2002). The assets the company requires for its innovation projects are its people, computing and prototypes. The company does not generally acquire assets from outside the organization, with the exception of the case listed above.

5.7 The Role of Sustainable Development in Building Innovation Capability

Sustainable development (SD) exists through superior waste minimization, recoverable manufacturing, supplier conformance to SD, recycling (of scrap steel, paper and cardboard) and engine technology to reduce emissions. In relation to engine emission reduction, SD is considered an innovation capability. UGM's strategy is to be technology leaders, which influences the organization to consider SD in its innovations. Its strategy includes a critical success factor to meet or exceed all appropriate government regulations:

UGM has a small number of environmental policies, reflecting the minor environmental impact of its facilities. Caterpillar leads the underground industry with low emission engines. Government regulations mandate emission levels for diesel engines in underground equipment. In addition, its parent company has a clear sustainable development statement in the corporate mission.

The sustainable development orientation (SDO) strategy was implemented through written waste management policies and implementation of these written policies, and also through environmental audits every 2 to 3 years by trained external auditors. In 2003, Caterpillar became the first engine manufacturer to offer a complete line of 2004 model year clean diesel engines that are fully compliant and certified by the US Environmental Protection Agency (EPA).

Caterpillar's breakthrough emissions control technology, known as Advanced Combustion Emissions Reduction Technology (ACERT), is designed to comply with EPA standards without sacrificing performance, reliability or fuel efficiency. Named on the Dow Jones Sustainability World Index in September 2002, Caterpillar is recognized for successful integration of long-term economic, environmental and social aspects into business strategies that benefit all stakeholders. Caterpillar's commitment to social responsibility ensures its ability to meet today's needs without sacrificing the ability to meet the needs of future generations.

Caterpillar continues to play an active "good global corporate citizen" role by working with organizations committed to making the world a better place — the Global Mining Initiative, Tropical Forest Foundation, World Business Council for Sustainable Development and others. Caterpillar believes that within the context of market-based environmental regulations and free trade, the business community can make critical contributions to a more sustainable world. In May 2004, it served as the sole industry supporter of the Global Mining Initiative conference, which provided a forum for stakeholders to express concerns about the past and future of mining.

Caterpillar is also an active member of the World Business Council for Sustainable Development, which provides business leadership as a catalyst for change toward SD and promotes the role of eco-efficiency, innovation and corporate social responsibility. Its on-going efforts relating to reusability of components, machine rebuildability, exhaust emissions, noise level improvements and safety enhancements were recognized as positive contributions. Caterpillar is also succeeding in making its facilities and equipment more environmentally friendly. The Illinois Environmental Protection Agency named Caterpillar's Mossville, Illinois facility the Best Operated Industrial Wastewater Treatment Plant — number one among 1,613 plants evaluated statewide.

An environmental management system (EMS) is part of UGM's overall management system. Specifically, it assists management and other staff to manage their operations in order to meet their environmental legal obligations, achieve corporate objectives, reduce wastage and avoid any unintended environmental impacts. An effective EMS will reduce pollution risks, the potential for environmental incidents and the risk of legal non-compliance, and avoid the costs associated with such incidents; enable the company to demonstrate due diligence in environmental issues; enhance public image and offer marketing opportunities; preempt any customer requirements for a certified EMS as a condition of business; and achieve cost savings through improved environmental performance and reduced wastage.

By integrating sustainable development philosophy and policies into its business model and developing new green products that meet strict US environmental standards, Caterpillar demonstrates a clear understanding of the importance of sustainable development and its positive impact on the organization's innovation capability.

5.8 The Role of e-Commerce in Building Innovation Capability

UGM has embraced e-Commerce as a new way of doing business. This is evident in the reduction of manual or paper-based systems. e-Commerce is considered as a means of cost reduction:

> Communication needs, customer demand and supplier demand are all external factors that influence the organization to use internet-based technology in its innovations. Some internal factors that influence the organization to use internet-based technology in its innovations are the company's strategy and cost reduction.

The main uses of the internet-based technology at UGM are the intranet, internet, funds transfer, parts order placement, company website and data transfer. UGM believes that information technology will drive machinery development in the 21st century. Caterpillar is at the cutting-edge, continually setting and improving benchmarks:

> I think e-Commerce certainly contributed to our innovation capability in terms of engineering, knowledge and sharing. Being a part of a larger organization made it much easier to tap into the other knowledge that exists in the other parts of Caterpillar. It's certainly changed the way we communicate, so within the space of five years, going from 90 percent paper to 90 percent emails, and certainly transferring knowledge and that sort of thing. In terms of commerce itself, we use e-Commerce in a number of areas. Most of our payments to suppliers are done now by e-Commerce. A lot of our order processes are now electronic — hand-written or faxed orders. We have not yet to this stage sold anything by e-Commerce, nor do we expect to, because large capital pieces of equipment are not something that people are just going to order over the internet.

Caterpillar is using sensors and transducers throughout key components of machines to continually monitor operating systems, machine productivity and availability, maintenance scheduling and fuel efficiency to drive lifecycle costing even lower. The next generation of Caterpillar machinery will utilize even more sophisticated systems such as combining Global Positioning System (GPS) and emerging technology, which will allow mining and farming machinery in the 21st century to become autonomous and semi-autonomous.

Application of technologies such as controled throttle shift transmissions in larger machines will provide huge gains in component life. Improved economies and reduced environmental impact will result from technological refinements in the new breed of electronic engines and other major component controls. Sophisticated and intelligent systems will link Caterpillar with its dealers and customers throughout the country. Wireless communications between a machine in the field and the Caterpillar network or the machine owner's maintenance facility can relay failure information instantly, thus cutting down-time.

5.9 The Role of New Product Development (NPD) in Building Innovation Capability

A strategic challenge facing UGM is how to accelerate the NPD process without undermining the quality of its output. This, however, is not a pressing issue at the moment because the firm is struggling to meet a backlog of orders, which is hampered by a national skills shortage, rising price and availability of steel, and the desire to retain existing customers. The organization has in fact had to slow down the NPD process to allow for a return on investment from innovations. The organization has been using Six Sigma as a platform for NPD as well as multi-functional teams to work on a rotation basis in the new product development area:

> New product development is accelerated in the organization by a disciplined NPD process, responsible and experienced staff and cross-functional steering committees. e-Commerce has contributed to the acceleration of NPD by way of data transfer over the internet. Sustainable development has no impact on NPD. Caterpillar sees its organization as having sustained differentiation advantage and cost advantage with respect to its product innovation from NPD.

UGM shares knowledge on NPD with other divisions of the global Caterpillar organization through the intranet and other networking opportunities:

> There's always interfaces between parts of the company and certainly we've tried to make sure that there's a fairly good rotation between those areas, so that one group isn't seen as elite. When we start a new project, we don't have people permanently, apart from the new management group, in the new product area. We start a new project and we'll move people out of other areas into a project team.

Management of Technology

UGM's focus on being the technological leader in its chosen fields and markets has meant that the organization has had to master the art of managing technology for innovation performance. The company has addressed this challenge by specializing in underground mining machinery and having a better understanding than its competitors of what the needs of its customers are in specific fields of operations and what is required to stay ahead of the pack. Continuous research, training and education have enabled employees to develop and maintain the necessary skills and motivation to manage the innovation process and to make the most of existing and emerging technologies.

5.10 Organizational Performance

UGM measures its organizational performance every quarter or every month against metrics for all critical success factors based on the balanced scorecard approach to determine if its strategy is working. The range of very specific metrics used to analyze the organizational performance as well as the performance of every project include financial, marketing, HR, manufacturing and NPD:

> When we talk about organizational performance, of course according to the balanced scorecard, we look at financials, customer, process and innovation. I guess you are doing all of those with key performance indicators.

UGM does not use triple bottom-line reporting — it does not see a need or benefit for this application at this stage. In terms of self-assessment and continuous improvement, the organization uses ISO 9000 Quality to comply with international and national standards. It is interesting to note that only one customer has queried whether UGM has ISO Quality Certification, indicating that branding is more important to customers than quality certification per se. Innovation is measured by commercial success, which depends on how customer-focused the organization is and whether it is coming up with innovations that customers want.

According to Lawson and Samson (2001), innovation performance can be measured based on revenue from new products, innovativeness, customer satisfaction, productivity, employee morale and

R&D as a percentage of total sales. The implications for the organization in terms of strategies for enhancing its innovation capabilities can be analyzed using the Lawson and Samson (2001) conceptual model illustrated in Fig. 4.1.

5.11 Leadership and Culture

Caterpillar has long understood the importance of leadership and strategy in enhancing its innovation capabilities. The leaders of UGM have formally included "innovation" as part of the company's vision and mission statement, and have provided training in Six Sigma and other techniques to facilitate positive innovation outcomes. An entrepreneurial culture conducive to innovation has helped create a climate where strategy is driven by customers' expressed and latent needs. The process of innovation has been championed from top-down and is part of the organization's key performance indicators. The company has worked closely with existing customers, suppliers and dealers to pilot-test its newstream strategies.

The competence base of the people working at UGM has a direct relationship to leadership competence, corporate performance and innovation. This is why the organization has invested in establishing and running Caterpillar University in the USA and a training center in Australia. UGM rotates people on different projects, thus enabling them to multi-skill and to have a better appreciation of how the system works with all its interconnected and interdependent parts. In this context, innovation performance and the competence base of the workforce appear to have a strong and positive relationship.

UGM has invested in new information technology and knowledge management as a means of reducing costs, improving productivity and communication, and facilitating innovation. The internet and intranet have been recognized as innovation channels for gathering and sharing information and converting it into marketable products. The case study highlights the importance of organizations having a market and customer orientation, and the positive impact that this can have on innovation performance. UGM actively and systematically seeks feedback from its customers and dealers in order to better align its products and services with the needs and expectations of customers. Having a strong focus on existing customers can be potentially dysfunctional if new customer segments and new markets are ignored. This creates opportunities for innovative entrepreneurial SMEs to carve a niche for their products. However, the high cost barriers of market entry into the underground mining machinery sector have prevented new players from entering the market.

Creativity and ideas management do not just happen by chance. Rather, they are a product of a system that encourages innovation and entrepreneurship and provides resources for people to engage in critical thinking. UGM's focus on creating and sustaining a global niche dominance in underground mining machinery has been supported by a system and an environment that nurtures creativity and ideas management. The integration of continuous customer feedback and market intelligence in the creation of innovative new products has been instrumental in facilitating continuous innovation at UGM. The promotion of a knowledge-sharing culture based on rotation of staff on diverse new product development projects and the use of information technology have enabled UGM to profit from creativity and ideas management.

Organizational structures and systems can help or hinder the innovation performance of the organization. UGM has a flat organizational structure, and its systems for supporting creativity and ideas management have played a positive role on the innovation performance of the organization, as measured by various innovation metrics.

Research has shown that culture and climate can have a decisive impact on an organization's innovation performance. The culture and climate of UGM have been influenced by the entrepreneurial and innovative approach of the parent company that creates conditions for creative thinking, risk-taking and a strong focus on markets and customers. The culture and climate for innovation and entrepreneurship have also been influenced by the Australian CEO and senior management, who recognize the importance of linkages between culture and climate and innovation performance. The use of Six Sigma and the balanced scorecard methods has also contributed to a culture of accountability, transparency and recognition of performance.

5.12 Lessons Learned and Opportunities for Improvement

The main lesson that UGM has learned from the current skilled labor shortage is that the continuous development of the competence base of the people working at UGM has a direct relationship to corporate performance and innovation capability. Secondly, rotating people on different projects enables them to multi-skill and to have a better appreciation of how the system works with all its interconnected and interdependent parts. In this context, innovation performance and the competence base of the workforce appear to have a strong and positive relationship to innovation capability building.

Caterpillar intends to remain focused on retaining leadership in the mining, construction and quarrying industries, as well as growing its business in agriculture, forestry and in the markets using paving products and compact utility machines. Customer research shows that there are growth opportunities in new engine-related products for the broad range of electrical power generation, petroleum, marine and industrial applications. Availability of parts and fast and efficient distribution through a network of independently-owned dealers has long been a hallmark of UGM's operations in Australia.

UGM's continuous learning culture will ensure that all support systems will be constantly up-dated through dealerships to customers. This begins internally where employee skills are regarded as an investment, not an expense. Continuous improvement programs cover technical as well as "people" skills, such as teamwork. These values go hand-in-hand with the company's vision and mission discussed earlier. These skills and improvements are passed on to the customer, through UGM and its dealers, as a better track record of delivering value — a commitment to the customer on lifecycle costing that is lower than any other competitor machine.

5.13 Conclusion

Based on the qualitative analysis, we articulate some key findings regarding the exploitation and use of innovation capability concepts, principles, tools and techniques and their relationship to innovation performance. UGM develops innovation capability through the simultaneous mix of leadership and strategy, people competence, information and organizational intelligence, market and customer orientation, creativity and knowledge management, organizational structures and systems, culture and climate, and management of technology. This results in an impressive innovation performance, as evident in the firm's global niche dominance. The company has successfully used innovation enablers like new product development, e-Commerce and sustainable development to boost its innovation performance.

Review Questions

(1) Explain why UGM faces a challenge in accelerating the New Product Development (NPD) process. How did UGM attempt to accelerate the NPD process? What is the likely impact as a result of a shorter NPD process?

(2) What are the key lessons learned from the UGM case study?

Chapter 6

Drivers of Innovation Capability at Sun Microsystems (SMS)

Milé Terziovski and Christopher Barnes

6.1 Introduction

Sun Microsystems (SMS) is a good example of what constitutes innovation capability in organizations, and how it can be developed and exploited. In 2005, SMS was a US$14 billion global information technology and communications (ITC) company, specializing in providing complex network computing. Driven by a culture of innovation, entrepreneurship and foresight, the classic Silicon Valley start-up company had grown from four employees in 1982 to 35,000 employees in 170 countries in 2005.

The case analysis has found that SMS's competitive advantage can be attributed to its vision, mission, customer focus, innovation, cooperation, commitment to quality service and making computer power more affordable. SMS has taken advantage of the innovation enablers like new product development, e-Commerce and sustainable development to continuously improve its innovation performance and sustain its innovation capability, which is the principal source of its competitive advantage.

The ITC industry is a growth industry characterized by constant innovation, expansion and application of technological output and reduction in the cost of hardware and software. It is an industry that has undergone rapid capitalization, a sharp stock market adjustment resulting from the "dotcom technology stock crash" and a return to some logic, following a period of investor exuberance. The future of firms competing in the ITC industry is determined by *inter alia* their innovation capabilities and how fast they develop and commercialize new products for existing and new markets. This provides a unique context for examining the innovation process and how it contributes to organizational performance using the case study method.

SMS is a good example of an innovation-driven firm that for the last two decades has been on the forefront of network computing. Information for the case study has been collected through an in-depth interview with a key respondent from SMS as well as from annual reports and material found on the company website.

6.2 Company Background

Sun Microsystems was incorporated in San Francisco in February 1982, with four employees. From its inception in 1982, all of SMS's systems included a network interface and all its employees were using electronic mail. Since then, SMS has established a history of innovation and leadership that stretches from the protocols that propel the internet to its widely adopted Java technology. SMS has provided innovation breakthroughs that have changed the way people work and the way companies do business.

According to company annual reports, SMS has helped companies leverage the power of the internet in virtually every field — from manufacturing to financial services; telecommunications to education; retail to government; energy to healthcare. The power of the internet allows companies to streamline processes and raise productivity, and reduce both costs and complexity. SMS has understood the critical issues customers face, and offers them quality products and comprehensive services that migrate the customer's business to a new level of competitive advantage.

As broad as SMS's product line is, the company recognizes that no single supplier of computing solutions can be "all things to all people". That is why SMS has established long-standing relationships with leading companies worldwide. These include companies with value-added resellers who add capabilities to SMS systems for use in specialized markets; original equipment manufacturers who incorporate SMS products and technologies in everything from embedded controllers to massively parallel supercomputers; and independent software vendors who work with it to deliver tuned and tested business solutions for enterprise resource planning, supply chain management, data warehousing, etc. Systems integration firms use SMS products to deliver comprehensive, integrated solutions for mixed-platform environments.

SMS helps design, test, deploy and manage network computing solutions that can translate into real competitive advantage. SMS's professional services experts have provided single-point-of-contact solutions to fit business needs such as managed services, utility computing, high-availability service packs, customer relationship management, network identity services, etc. SMS is also committed to fast response. Its world-class team provides mission-critical support 24 hours a day, seven days a week. It also provides multi-vendor support that solves complex problems with one phone call.

SMS employees are its most important resource and the basis for its success. The corporate culture can be described as an environment characterized by respect for each individual, where cultural and ethnic diversity are blended by teamwork into a harmonious workforce. In Australia, the company's respondent to the in-depth interview was Mr Chris Barnes, Clients Solutions Executive. The respondent is part of a business unit called Client Solutions and has 8,000 people internationally. The organization has 35,000 employees in total. The structure of the organization is hierarchical, but it also operates in a matrix mode.

SMS has access to the latest technology with video through the internet. In the past, SMS interacted via telephone conference for meetings, then video conference and now web conference, which is a combination of video conference, phone conference and internet podcasts.

6.3 Corporate Strategy

SMS's vision is to provide highly scalable and reliable ITC systems that will connect everyone and everything to the network. In virtually every sector of the economy, SMS is helping organizations leverage the power of the internet. Continuous product innovation such as the

Java technology has enabled developers to write applications to run on any computer. A fundamental element of SMS's strategy has been reinvesting a significant portion of revenue back into R&D in an effort to maintain its performance advantage (Shafer *et al.*, 2005). Up to 2002, SMS enjoyed considerable success by avoiding the industry trend towards standardized chips and software (Tam, 2003). SMS's strategy was to offer more powerful and more expensive computer solutions based on proprietary hardware and software, which worked well as long as it was able to maintain a performance advantage. However, standardized chips eventually matched the performance of SMS's proprietary chips and standardized software offered functionality similar to SMS's. As a result, SMS had seen its quarterly sales drop by more than 40 percent since its peak in 2002, and its stock price decline to under $4 per share from a high of over $60 per share. The strategic choice made by SMS in late 2002 to add a line of cheaper servers based on Intel chips had a significant impact on the firm's ability to maintain its current R&D spending levels, which in turn had implications regarding its ability to compete on the basis of higher performing solutions.

6.4 Mission Statement

The word "innovation" does not appear in the mission statement, but it is part of SMS's culture, part of its aspiration to be the innovation leader in technology. SMS defines innovation as the ability to be successful in using technology to optimal advantage for the organization:

> For our business unit, "innovation" means that we need to focus on the ability to successfully deploy our products in the most cost-effective way for our clients. It means we need to update our methods of deployment, for example delivery, which are constantly challenging.

Five years ago, SMS was using a waterfall approach for system development. Today, it follows a totally different approach in terms of software delivery. SMS has had to keep changing its processes in conjunction with supply chain partners to bring ideas to market.

6.5 Core Competencies

SMS has a number of core competencies that have enabled it to survive and thrive in network computing since it was established two decades ago. The first relates to its ability to continuously innovate by investing one-fifth of its sales revenue in new product development. The second core competence lies in niche technologies such as Java, Unix and Identity Solutions:

> One of the biggest issues the corporate world and government are facing is authentication of identity. Our product enables you to pass from one system to another and seamlessly manage and maintain that security. We are still in a formative stage and we can exploit this core competence internationally.

6.6 Innovation Strategy

The literature shows that a range of external and internal factors influence innovation performance in organizations (Metz *et al.*, 2004). External factors include government regulation, environmental regulation, e-Commerce regulation, customers and competitors as well as partnerships. Internal factors that have been empirically proven to influence innovation include organizational size, strategy, structure, type of organization, slack resources, culture and climate, communication, social structures, people and HR management, management and leadership, knowledge management, teams, incentives and rewards, management of technology and market knowledge.

SMS clearly understands the importance of continuous innovation and invests 20 percent of its revenue in R&D:

> A lot of companies are at less than 2 percent and we are at more than 20 percent.

The industry average is less than 5 percent. SMS can afford to invest a large amount of its revenue on R&D because the history of the company shows that its growth is attributable to heavy investment in R&D. For example, the investment in the innovation called "Driver" catapulted the company as did investments in Sparc Solaris. Hence, its major new innovations in the industry are directly attributable back to its investment program in R&D.

At the time of the interview, SMS was in the process of announcing Solaris 10, which is a total revolution in operating systems and a direct result of the firm's R&D efforts. SMS has a wide customer base. Every organization is a potential customer if it uses a multi-platform to run SMS equipment. SMS differentiates itself from other organizations in the industry by taking the approach to minimize risk for the client. SMS prefers to be in a situation where it is unique and the only firm that can deliver that service:

> In the same way we've got the open theme in our products, we actually have the open theme in our service delivery ... so that we can leverage a partnering network to actually deliver Sun services rather than having Sun deliver in the same way that IBM Global Sevices tries to deliver most services themselves. We want to go the other way so that Sun or any of its partners can deliver Sun solutions.

The products are designed so that customers can plug and play with other third-party products as long as the system has some component of SMS. SMS has filled in a niche by being the so-called "plumber of the industry". SMS measures its ability to continuously innovate by using Six Sigma processes. It has a process called Return on Sun Sigma, i.e., the number of dollars or revenue that is generated against the investment it puts in. This is measured through their systems. SMS is introducing a lot of patented new products on the market and measures its performance through benchmarking:

> We benchmark and measure against our competitors in terms of market share.

In terms of the breadth of resources within the field of Client Solutions, SMS elects to go with specialists to architect and manage risk in delivery of complex solutions. Resources in Client Solution are pre-sales and half of them are delivery. A small group in each vertical for each product does all the thinking in terms of product planning.

SMS also has an "engine room" that looks at process and how to match that up with the products:

> We are looking at how to push out to market solutions, not products.

6.7   Resource Availability and Absorptive Capacity

The availability of slack capital and resources has been identified in the literature as a prerequisite for successful new product innovation. SMS draws on its extensive internal and external resources for creating and capturing value within its value network. One of the biggest challenges facing SMS is maintaining its ability to continuously absorb and act on information and knowledge from the external environment:

> It's an area that you have to work very hard on, and it's an external process and your natural inclination is on internal process. It is something you consciously have to do and that is why we have Sigma Black Belts.

The strategic information generated by Six Sigma Black Belts is used to drive the sales process. In order to maintain the innovation drive, SMS invests in assets. It has good facilities in terms of having access to laboratories and work centers.

According to the respondent, the organization's most important asset are its people. Getting commitment from the right human resources is the biggest issue. There appears to be a significant and positive relationship between the quality of the people and quality of outcomes. SMS acquires assets from outside the organization to enhance its innovation capability. In particular, SMS partners with other companies in its industry. To put together a solution, managers at SMS often need other bits and pieces which are not SMS components, so they bring those people in as partners.

SMS drives and implements projects with the goal of creating a win-win situation for the customers. Occasionally, some of the other partners have an idea which they are running and SMS helps them with facilities. Sometimes, a customer may require SMS's assistance with a go-to-market strategy. SMS has six Global Solution Centers, including one in Sydney, which are innovation environments where customers, partners and SMS can put together a solutions testamur and use it for marketing. A project may run for three months with customers in order to be able to establish whether it works.

Customers have reported savings of $300,000 in time and money by being able to use this facility to test out their innovation. SMS uses a two-plus-two approach to absorb information on innovation from the external environment. Generally, SMS goes through a qualification process which consists of a two-hour presentation to see if an idea gets some traction. If the idea gets some traction, SMS will do a two-day workshop and flash out a proof of concept. From that point, SMS will do a two-month or two-year proof of concept, whatever it turns out to be. It is taken in stages and is a spaced approach so SMS can manage risk and also get the buy-in of stakeholders. The knowledge is stored in SMS's extensive work sites and systems. Knowledge-based centers have actually been built to service its needs on a global basis so that it can leverage the benefits of being a global organization.

6.8 Innovation Capability

Innovation capability has been defined by Lawson and Samson (2001) as the ability to continuously transform knowledge and ideas into new products, processes and systems for the benefit of the firm and its stakeholders. Sustainable development (SD) is part of the company's philosophy and management practices, with a focus on product "after use". In particular, how a product will be disposed of, recycled or re-exported is an important part of the product policy mix. SMS follows a ten-point sustainability criteria which includes the requirement for the customer to be non-hostile to the US. Furthermore, SMS has also recently revamped its workplace and work-from-home policy. The company has revamped all of its offices on a global basis in order to save energy and accommodation space by better utilization. Most knowledge workers at SMS work on a flexible basis and they book a workstation as required, rather than holding a permanent space at a site. This innovation has saved SMS about 30 percent in its worldwide cost of accommodating staff.

In many respects, SMS is very much a virtual organization that is influenced by both external and internal forces. The key external factors that influence the organization are government regulations, competitors and customers. SMS's sophisticated and demanding customers focus on value and how to increase their own competitiveness. The key internal factors that influence SMS's SD practices have been its CEO and senior management, who have a very strong vision for the organization. These internal factors drive the R&D program and decisions concerning the firm's environmental performance. SD considerations are also reflected in the organization's culture by way of shared values and fostering an environment where the people within the organization are encouraged to be innovative. The organization offers formal and informal rewards for innovation contributions. Anybody in the organization whose idea is adopted and shows return receives a reward for their efforts, both intrinsically and extrinsically, in the form of financial rewards and also recognition and part of performance review:

> The organization has an incentive reward program that can be given on a stock/stop basis.

Some examples of the implementation of SD innovations within the organization are in the office environment. Physical layout is a clear example. The policy of encouraging digital storage as a way of minimizing paper usage has also been successful. Each SD initiative is implemented in a planned way. SMS has a measurable component where every internal process or initiative undertaken has a return on Six Sigma. Each division uses Six Sigma methodology with appropriately trained Black Belts and Green Belts who contribute to the innovativeness of the organization.

Throughout the organization, SD is considered an innovation capability. The organization relies on the innovation capability of its people, their mindset and their ability to deliver products that are going to be recyclable and have zero carbon emission:

> We want that same commitment to sustainable development to exist through our channels, through our partner community.

SD permeates through the whole organization and is manifested in the organization culture. SMS is a global leader in enterprise computing and is committed to providing timely and useful information regarding the company's impact on the environment, its programs, and the health and safety of its employees. The corporation has signed on as a partner with the Environmental Protection Agency's Climate Leaders Program. SMS recognizes the importance of natural resource conservation. The company implements conservation technologies in both new and existing facilities, thus helping the environment by limiting the use of precious resources and saving money. SMS's recycling programs consistently divert tons of materials from landfills and save literally hundreds of thousands of dollars per year. Its recycling and solid waste reduction programs include waste reduction, office paper recycling, pallet recycling, cardboard recycling, foam recycling and computer equipment recycling.

SMS has initiated and funded several clean air and alternate commuting programs to reduce significantly the adverse environmental aspects of commuting. An important part of SMS's commitment to responsibly manage all aspects of its business is minimizing the environmental impact of its products while still maintaining the extremely high standards of reliability and availability. Its Design for the Environment program assists product engineers in developing products and packaging designs that incorporate supplier management, energy efficiency and design for recovery, reuse and recycling.

6.9 The Role of e-Commerce in Building Innovation Capability

The internet has forced some organizations to review existing processes and practices and to reconfigure their innovation capabilities. SMS is a good example of how internet-based technology has influenced the innovation capability of the firm. Internet-based technology is part of SMS's strategy in the way employees work and how the organization is virtually structured. Furthermore, the intranet has provided flexibility, security and the ability to retain knowledge. SMS has put a significant investment in its intranet systems. This has taken a lot of cost out, but has built a lot more efficiency in:

> It is easier if your company is in the dotcom scenario. If you've got a start-up business, it is much easier to make a new business efficient than it is an existing business. You can set the culture right from the outset rather than trying to change it. You don't have legacy systems to worry about, you don't have existing infrastructure, and you don't have to worry about cost benefits and changing those. So that's certainly a factor for us.

Given that everything in the organization is internet-based, e-Commerce was quickly embraced. Around 80 percent of SMS's business comes from large corporate bodies and governments who are mature e-Commerce users. e-Commerce enables SMS to extend its business model and to change the effectiveness of its business model. Its hardware is ten times exponentially increasing in sales, but the costs are coming down:

> If we stay in the hardware business only, we are going to go out of business.

e-Business enables SMS to extend its range of services and thus to position itself as solving customers' problems. The evidence from the literature indicates that most organizations around the world have yet to develop competencies in e-Commerce and SD practices with traditional business models and approaches (Dunphy *et al.*, 2003). The SMS case study shows that both e-Commerce and sustainable development are contributing factors within innovation capability.

6.10 The Role of New Product Development (NPD) in Building Innovation Capability

The literature shows that the process of accelerated NPD is considered to be increasingly critical to the competitiveness of organizations (e.g., Mabert *et al.*, 1992; Pawar and Sharifi, 1997). The case study shows that SMS has successfully used innovation centers that focus on accelerating NPD in their organization. The innovation centers are made up of business people who are part of teams focused on continuous leveraging. The organization has yet to take advantage of e-Aim, which is a knowledge-capturing innovation that can have a significant impact on the acceleration of NPD. e-Aim can assist with the process of developing NPD by allowing people who are working on different project teams or innovation teams to share knowledge and learn from other parts of the organization, and extend this through SMS's network and supply chain. Like in other innovation-driven organizations, e-Commerce and SD practices have facilitated the acceleration of NPD at SMS. Combining the two together has resulted in more innovation capability:

The net result is lean and green.

There is only one instance where sustainable development and e-Commerce may hinder NPD acceleration, and that is during passing the test in regards to return:

If you say as a result of those constraints you are not going to pass that test and you're not going to get a return — and the return can be measured in multiple ways, not just dollars — if you're not going to get a net benefit, then you are not going to do it.

SMS has developed cost advantage with respect to its process innovation for NPD relative to other traditional methods. However, it has yet to develop cost advantage relative to its competitors. SMS provides a range of hardware, software and services to support the technology vision. It is based on open standards so that everything it does has an open, non-proprietary approach to it in the context of its fit in the marketplace, so that it can interconnect with other organizations' products to provide flexibility and leverage on existing investment of an organization's infrastructure.

SMS has a strategy for NPD and a dedicated department for innovation. The company makes use of cross-functional teams as part of the NPD process. Customers and suppliers are regularly consulted on how SMS products can better meet their needs. Branding plays an important role in facilitating the introduction of new products onto local and international markets:

> If you have a look at the history of the company and growth of the company, it has been directly attributable back to the continual investment into R&D.

6.11 Organizational Performance

Some parts of the organization have adopted a balanced scorecard approach, while others have adopted parts of it but not the full picture. This is the case with SMS in Australia. All parts of the organization have a set of metrics which get measured very stringently. Checks and balances are done on a weekly basis. Innovation management, together with financial performance, is the main focus. The application of the triple bottom-line reporting has begun in Australia, given that SMS measures environmental and social performance as well as financial data. It also undergoes self-auditing twice a year. The information is recorded in a financial template to meet the financial standards. SMS also has process templates to ensure that everything goes in the right sequence and are signed off by the right people, which then becomes supporting documentation.

6.12 Opportunities for Improvement

In terms of competitive advantage, SMS has not sustained differentiation advantage with respect to product innovation from NPD. The respondent was of the view that the organization is still not quite there yet in terms of fast product innovation. Furthermore, in terms of cost advantage, SMS has yet to have maximum cost advantage relative to its competitors. The two largest areas of improvement are supply chain and delivery process:

> There is also a focus on marketing. We work on improving our progress with respect to marketing and get a better return out of that.

6.13 Conclusion

Using the conceptual framework of the Innovation Capabilities Model (Metz *et al.*, 2004), this case study examines the influence of the drivers of innovation such as sustainable development, e-Commerce and new product development at SMS. Based on the interview and supporting literature, it can be concluded that these three elements facilitate the innovation performance of the organization. This finding supports the Integrated Innovation Capability Model and its hypothesized linkages.

Furthermore, the case study shows that innovation capability can be built on or enhanced through expertise gained in NPD, e-Business and sustainable development. A number of implications for management of innovation performance can be observed from the analysis. First, the innovation capabilities of the firm are an important driver of innovativeness. For example, SMS's strategy, absorptive capacity, slack resources and human resource policies have a direct impact on its innovation performance. Accordingly, managers need to understand the interconnected nature of the innovation process and how each component of the firm's innovation capability can influence innovation performance.

Second, the innovation performance is influenced by the quality and intensity of SD practices which help the organization meet the expectations of government regulators, customers and suppliers. SMS has managed to achieve significant cost savings by eliminating waste from its design and production processes and by moving towards digital storage of information. Third, the qualitative case study shows that SMS has improved its innovativeness by making full use of e-Commerce to meet the needs of internal and external customers. Finally, NPD has been accelerated by innovative management practices such as Six Sigma, which has enabled SMS to focus on cost drivers, improve service and ensure sustainability of its continuous improvement strategy. The case study demonstrates that innovation capability provides the potential for effective innovation and gives valuable practitioner guidelines.

Review Questions

(1) Discuss the role of e-Commerce, sustainable development orientation and NPD in building innovation capability at SMS.

(2) Why do you think SMS had a challenge maintaining its ability to continuously absorb and act on information and knowledge from the external environment?

Chapter 7

Development and Exploitation of Innovation Capability at a Defence Project Engineering Company (DPEC)

Susu Nousala and Milé Terziovski

7.1 Introduction

As Australia attempts to sustain its global competitiveness, several studies of how firms in Australia achieve excellent performance in product innovation have been gaining much attention (Terziovski *et al.*, 2002; Gloet & Terziovski, 2002). These studies suggest that firms can increase their competitive advantages and capture innovative knowledge by accelerated and shortened product development cycles. For firms to achieve this, various integrated expertise in NPD, e-Commerce and SD are needed to successfully transform innovative capability into innovative product. This chapter investigates the elements and processes of building innovation capability in the Defence Project Engineering Company (DPEC) using an innovation capability project survey protocol, specifically designed for the development of the case study. This case study identifies innovation practice and highlights the competitive gaps which could contribute to future improvements in innovation (Clark & Fujimoto, 1991).

7.2 Company and Industry Background

DPEC is a subsidiary company of its parent company. It is a large project management and technology contractor, employing 4,000 people, of whom several hundred are engineers. The company was established over five decades ago and began as a single-project company, and over time began diversifying into other projects and sectors. During this time, the company also diversified into the defence sector, establishing diversification and capability within this sector, including lead engineering and maintenance contracts. The company, at this point, successfully carried out multiple major complex engineering projects, both nationally and internationally.

Since 2005, the company has provided project management services to major private infrastructure projects. DPEC continues its extensive research and development into the commercial application of products, services and processes through the commercialization of its technology. DPEC assets include defence and project management businesses, infrastructure maintenance and engineering services, as well as high technology ventures with European firms and US.

The company DPEC is comprised of six main divisions, which cover areas such as design, manufacturing, modifications and repairs to military vehicles with appropriate quality accreditation. Military systems are supported by facilities in engineering, integrated logistic support, through-life support, configuration management, prototyping, warehousing and production. Certain divisions also provide logistic support, which involves warehousing and maintenance of diverse military systems. Other divisions specialize in system solutions with regards to command, control, communications and intelligence, surveillance and reconnaissance, electronic warfare, simulation and hydrography. The specialist capabilities include systems engineering and integration, design, development, installation, test, manufacture and all elements of integrated logistics support. Divisions also provide R&D for data and information processing as well as software.

7.3   Company Characteristics

The organization is primarily hierarchical and has autonomous business units within the divisions. Interaction occurs through various functional areas; for example, project management skills, finance skills and R&D management are carried out across all the divisions. There are also engineer managers within the divisions who exchange information about their functions. In terms of priority, it is the division itself that takes priority.

7.4 Perception and Definition of Innovation

The perception DPEC has of innovation is reflected in a definition offered by Porter and Stern (1999), which states,

> Innovation is the transformation of knowledge into new products, processes and services. It involves more than just science and technology — it involves discerning and meeting the needs of the customers.

For DPEC, innovation exists as part of a standard procedure embedded in product development, which eventually leads to the development of a product for its customers. Therefore, if the product is successfully developed and delivered, the innovation process has also been successfully developed and delivered.

7.5 R&D Department

The main role and responsibility of the R&D Manager of DPEC is to ensure that support and the appropriate resources are available for product development for all the divisions. There are three main roles for which the R&D Manager is responsible. These key roles demonstrate the suitability for the study. They also show that through these roles, the R&D Manager can facilitate activities which support the company in world-class projects, products and services. These roles are listed as follows:

- Develop new and improved products, processes and services with the objective of winning major and minor defence projects;
- Management and protection of intellectual property, ensuring/ identifying patents and trademarks; ensuring various contractual agreements are managed and engineers are in compliance;
- Leveraging R&D funds; federal, state governments.

7.6 Corporate Strategy

The company's vision and innovation is part of the corporate governance, which in turn is part of the company's core values — culture,
values, principles, management structure and company policies, procedures and processes. The R&D Manager articulated DPEC's purpose as follows:

> We're not here to do a project and shut down, we're here to try and
> expand and be the best in Australia.

The system of corporate governance is a key element shaping the
innovation process in advanced industrialized nations (Porter, 2001;
Hall & Soskice, 2001). Corporate governance is the system by which
large organizations are governed and monitored by specifying the distribution of rights and responsibilities among the different actors
inside and outside the corporation. The systems analyst at DPEC
elaborated on the importance of innovation there by stating that

> ... it's certainly brought out in our governance.

Innovation is espoused by DPEC through the company's mission
statement. The company's approach towards innovation and its mission statement is in line with the previous definition of innovation by
Porter and Stern (1999). The company's values are strongly related
to the corporate governance and mission statement, and are linked to
its innovation capability coupled with the corporate strategy and governance. There is a clear indication that DPEC has an innovation culture that is aligned with its corporate strategy. In the remainder of this
case study, we explore how innovation capability is developed and
how knowledge is transferred in the product innovation process.

7.7 Organizational Innovation Capability

The process through which innovation capability is viewed and developed at DPEC is through its team structure and the teams' environment in relation to e-Commerce, NPD and SD. The terms "teamwork"

and "leadership" are cited in the company's corporate governance and values, and reflects how the company perceives innovation, its improvements and its implementation. DPEC primarily utilizes small divisional teams (SDTs) to conduct its work within a project framework. These teams are often made up of the same people with overlapping responsibilities.

These teams could be seen as the building blocks of the projects within the divisions and are significant with regards to innovation capability. Fig. 7.1 shows what currently occurs with regards to

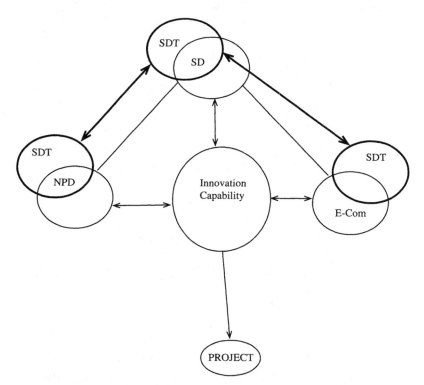

NPD, SD and e-Commerce support and create Innovation Capability for the organization and are organizational capabilities.

There is tension between the way the SDTs do work vs. the way the Project work is carried out.

Fig. 7.1 Innovation model — Small divisional teams (SDTs).

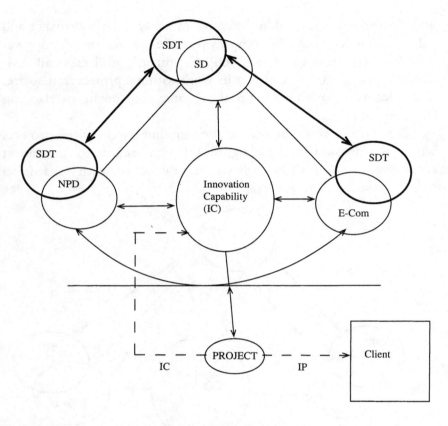

The SDTs create and support Innovation Capability through their ability to work across all necessary areas within the organizational divisions. This is due to the SDTs' ability to be flexible (being small and dynamic) with many individual members interchanging within and across teams as required. This model shows the possible tensions but also the importance of the SDT.

Fig. 7.2 Innovation model — SDTs, IC and IP.

innovation capability in the areas of e-Commerce, SD and NPD at DPEC (specifically in relation to the generic Integrated Innovation Capability Model (IICM) shown in Fig. 2.1). In Fig. 7.2, the model shows the impact of the work of small divisional teams and how these areas interrelate within the company. Fig. 7.2 also shows intellectual capital (IC) and intellectual property (IP) in relation to SDTs and the innovation capability areas during project development and product delivery to the customer.

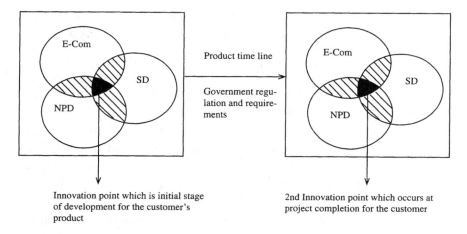

Innovation point which is initial stage
of development for the customer's
product

2nd Innovation point which occurs at
project completion for the customer

Fig. 7.3 Innovation points model.

Fig. 7.3 shows how innovation points occur. The three different areas of each circle within each box represent how the SDTs integrate with the relevant areas within the company. Not all of the relevant areas are absorbed or integrated into the innovation process, via the SDTs. The white areas show no integration, the striped areas show partial team integration and the black area in the center represents integration of all relevant teams and areas, where an innovation capability point occurs. The first innovation point takes place at the pre-contract signing stage, prior to the defence contract project deployment process being locked in. The R&D Manager explained how it could be possible for innovation capability points to emerge in the above model by stating that

> ... detailed design basically takes place before the contract; therefore there is very much less innovation after the contract is signed. DPEC needs to meet a specific performance specification .... The teams may want to do different things, but they get much more straight-jacketed after the contract is signed. But they are the same people going through the same problems with this structure ...

The first innovation point in Fig. 7.3 shows the initial design creation and problem-solving. The product timeline shows the product progress as the project moves through the various stages, so that it can be completed on budget and on time, according to the contract, regulation and compliancy requirements. The R&D Manager stated,

> ... as in fit for purpose and safe to use.

The second innovation point shows how the innovative knowhow can be embedded in the product when it is delivered to the customer. The possibility of retaining any innovative IP or IC after delivery is reduced for the company if it is not acquired before delivery at the second innovation point. It is important for the company to acquire the IP and IC at the second innovation point, whilst the original teams or individuals are still together. At this point, the original teams need to convert their know-how of applications and implementation of innovative products and processes, via an appropriate pathway, into "lessons learnt" for company-wide consideration.

7.8 Manufacturing and Services

DPEC has manufactured products through new product development (NPD) with its extensive services, which includes new service development (NSD). The effects of structure and strategy with regards to effective teamwork (which, in turn, supported the company's activities) and the effects on innovation are discussed by Atuahene-Gima (1996). In a study comparing innovation activities of Australian service and manufacturing firms, they found that NSD and NPD required differing support and organizational environments to succeed. Successful services within firms rely on employees (with effective self-management skills) in development and/or management to work directly with the customer.

These potentially long-term relationships with regards to successful services are less definable (or intangible) than the manufacturing aspects of a company. Unlike services within the company, quality, structures and procedures can be more specific and uniformly applied. Atuahene-Gima (1996, p. 37) states,

> Unlike a product, therefore, relationship quality is a more critical factor in success of a service .... Performance implications of the

differences between NSD and NPD activities draw on the strategy-environment co-alignment framework. This framework suggests that a fit between organizational strategy and the internal and external environments is a significant determinant of performance.

This framework of internal/external environments is significant not only for performance, but innovation capability itself. A strategy-environment framework would also be significant with regards to NPD, SD and e-Commerce.

7.9 Sustainable Development (SD)

DPEC is active in taking care of the environment as well as integrating into the community, whilst keeping itself in business. The company has a strong philosophy of performance in all areas, externally as well as internally (as reflected in its corporate governance). Part of the SD strategy was reflected in an example of core competency development, through the organization's waste management system, where the company developed a system to integrate throughout the organization, thereby ensuring that its suppliers were certified as well.

Internally, the organization's corporate governance covers SD. Externally, SD is primarily influenced by government regulations, which cover the defence sector and is generally considered more stringent and demanding than in many other industries. The company does not have an explicit corporate strategy on SD; however, many of the relevant issues are covered by occupational health and safety rules, which are adhered to as explained by the systems analyst in the following statement:

> ... I would say DPEC has a stronger philosophy of performance than the others .... [R&D Manager further confirms the stringent nature of the industry] In our case, it's not so much our customers or suppliers that are demanding, it's really government regulations which are demanding ...

DPEC takes into consideration community expectations, which also reflect government regulations. However, regardless of the government regulations, the company has always actively taken part in considering community expectations and has been involved in many community consultations. The company has actively succeeded in aligning itself with community standards, for example, company facilities or space used for community activities from time to time. As a result, the company is a welcome member of the area and community, which also reflects an alignment with the company's distinctive core values.

DPEC's corporate responsibility and approach to the local community was confirmed by the systems analyst:

> For example, you talk to anyone here about the possibility of closing down and they're not happy. They're happy to have us here .... There are a lot of things that the company tries to do to fit well into the community .... As a company we try to work in a way that conforms with community standards.

7.10 e-Commerce (e-Communication)

External influences for DPEC were mainly the use of the organizational internet-based technology in its innovation with customers, industry partners and suppliers. For example, the internet and intranet are used by teams to do research, product comparisons and generally communicate ideas, concepts and information. The speed at which information is exchanged contributes to defining a niche market space. For example, the company was able to commercialize technology which evolved from exchanges between teams of external research and technology organizations. This has allowed the company to upstage its competitors in competitive niche areas by adopting superior technology and usage, which again comes back to "project know-how and knowing how to make that effort", according to the R&D Manager. The R&D Manager further commented that

> It's interesting how quickly you can get information off the internet. We were looking at technologies for condition-monitoring through google, looking at websites so quick, and did it during a meeting. Google is also used for getting information outside the company for communication, sharing information and so on.

The internal influence for adopting internet-based technology has been the cost efficiencies. The selection of these technologies has contributed to a particular organizational culture towards technology, for example, ease of adaptability to the new technology and how it is seen to integrate with the current way of doing things.

7.11 New Product Development (Through Project Management)

The main factors that accelerate NPD in the organization are: clearly stipulated goals, availability of resources and improving their availability. With regard to the adoption of standard management strategies (such as QFD, TQM, etc.), the organization focuses on excellent product quality efficiency, which is project-driven. The organization certainly uses e-Communication to help product development in terms of information access and improved communication. With regards to the impact of SD on NPD, the impact was fairly neutral and was in line with required regulations and compliancy. The R&D Manager stated,

> ... if any management strategies were needed, then they were implemented.

DPEC is a project-focused organization, and as such has a system which drives its management to have a project focus, with a view to successful completion of the product for the customer. The following is an example of how innovation capability occurs under project conditions.

A decision is made to develop a product for a multi-million dollar project. This involves groups of designers (architects, electronic systems engineers) who put forward designs to win this project, and requires R&D funding support to do the work. Once the successful project is secured, the proposed design is developed into a prototype under contract, undergoing engineering analysis as specified in the contract. Much less innovation occurs during this stage since once the contract has commenced, the process needs to meet specific performance specification, etc. They cannot be brainstorming anymore, hence the process becomes more explicit. The same people carry out the initial (and often more innovative) work, contract work and final delivery. This allows for extensive communication for those in the SDTs to know what worked and what did not, the "lessons learnt" and subsequently, how to apply what was learnt.

7.12 Self-Assessment and Continuous Improvement

The several hundred or so engineers in DPEC spend part of their time focused on the three previously mentioned roles (i.e., new product and improvement of processes and services, managing intellectual property, leveraging funds for R&D). Within these roles, there are varying degrees of innovation. Products and services make up the defence systems and support products built by DPEC. To summarize defence systems, the R&D Manager commented,

> It could be described as a ship, patrol boat, destroyer, frigate or an armored vehicle, or the upgrade of an electronic system to an armored vehicle or aircraft, or military vehicle maintenance and repair, right through to life support and data management.

Products and services are adapted and used in other ways, which could be seen as value-adding, maintenance, an innovative process or a variation of an innovation process. A practical example of an innovation variation would be the development of a communication system. At the project's inception, the communication system was a new system that had involved a high degree of innovation during the systems development and production. However, after several years of operating, the system needed upgrading. This involved conceptual and detailed designs for NPD, a whole new product system and integration of existing key components of the design to create a complete design. After the innovative variation had been tried and tested, it was embedded as routine support work with innovative input.

As previously discussed, another aspect of innovation capability would be how the organization's competitive best product innovation is supported by relevant process innovation, i.e., the way in which you do business. Competing and acquiring work in the support areas is also relevant and involves process innovation. The organization's innovation capability can be seen with its technological development involving a vast transformation from a very manual approach of using sound to using laser technology to create a new product. According to the R&D Manager,

> It's a vast improvement, perhaps in the order of magnitude of a 100 times faster than the old system.

Another area of innovation capability is data security, through specific new data technology, which allows the linking of military classified network to unclassified networks such as the internet. This means that a person sitting in a secure area can see data in the unclassified area without allowing anyone else in the unclassified area to have any access to the military side. This technology was applied following certification of various standards, for example, the security stance for NATO, US and Australia. These standards are important as the company can then guarantee that hackers cannot get around the various barriers. There are other cases where DPEC have used physical barriers or separation on the same computer with associated hardware and software to guarantee security. This is the world's best product niche area.

7.13 Key Lessons Learnt

Although the company is project-based, lessons learnt were predominantly acquired through SDTs within their project parameters. The SDTs had variations of skill depth, depending on what was required. Typically, the organization uses small teams covering several business areas. Each line of military, electronics, etc., also has their own capabilities. While not specifically charged with being innovative (they are really charged with providing solutions within their R&D budget and timelines), innovative strategies did emerge out of solution-focused approaches, because by working as focused teams, they knew their own work and each others'.

An example of an innovative emergent strategy would be the laser product and the specific data technology, which evolved out of a relationship with defence science and technology organizations involved with the project teams. These systems are recognized as being world-class, and are the only systems in regular commercial service. The product undergoes continuous development so as to remain a leader in that particular niche product area. As a result of securing a niche in the market space, the product is currently exported to America, Alaska, Europe and the Middle East.

The strength and dynamics of these SDTs need to be expanded and perhaps more actively understood with regards to the development of project/business structural innovation. An example of this type of innovation came about through a major tender for product replacement. Serious competition ensued between DPEC and several other companies, who proposed different solutions to the tender for product replacement. A competitor won the project, with DPEC coming in second.

However, the interesting sequel to the DPEC design was that the very same design was successful in a project tender overseas. The overseas project required several vessels for protection and military transport. The success of securing the overseas project tender was due to DPEC's ability to localize the work and carry out a majority of the work *in situ*. This involved building the very same product design (that had previously come second in the tender process), which was now successful due to local previous project knowledge being carried out *in situ*. The project success was not due to product innovation, but had more to do with prior industry involvement, which was more of a business process innovation.

Being able to offer the final product within the overseas local economy (*in situ*) meant that AUS\$300 million from a AUS\$500 million project circulated within the local economy. Competitors could not offer the same localized capability that DPEC could provide. DPEC was then able to spin-off similar designs, which were licensed and successfully used again, against international competition. The R&D Manager elaborated thus on the lessons learnt from the company's success:

> DPEC ... builds a lot of this product in the overseas local economy, so the very same product design that lost before is now being built

in situ, overseas, not due to product innovation but prior industry involvement, which is a process innovation. It's nothing to do with technology. It requires our effort and our expertise, so it's more of a case that we have done it before with these big projects and we know how to make that effort.

7.14 Organizational Performance

Although the company process is driven by financial goals and the bottom line (which are seen as paramount), customers apply organizational performance and progress measurements to the organization. These are key criteria of the process/project performance. Organizational performance can and does improve through lessons learnt from previously completed projects. The level of impact that innovation capability has had on the organization is considered significant. However, due to the project-type structure, it is very difficult to separate from the project and apply any measurement. Success is often measured by the success of the project as a whole, and as such, innovation at that point is considered as a given because the work has been carried out. The focus is more on positive project outcomes, and not so much on performance parameters as such. Positive project outcomes define positive performance measurements or evaluation. This also reflects the way in which SDTs and the members carry out the work.

Although the previous example of the overseas project success could be seen as an innovation process (supported by continuous improvement via other concurrent projects), this continuous ability across the organization was not measured and is typically not a focused activity in itself. This is due to the project-type structure of the organization. The R&D Manager had this to say on measuring innovation within the project framework:

The project will be measured on its success.

The company has produced a number of patents over the last five years. In terms of project budgets, the awareness and close monitoring of what is spent for each project is rigorously measured. The R&D

component of a project is relatively small in comparison to the project budget, and as such is not measured in isolation.

R&D is funded and allocated via a standard budgeting technique. There is an annual update linked to a technology strategy, which is driven by what project the company wants to win, and a business strategy. The R&D Manager elaborated on these strategies:

> We have a technology strategy ... which says what the division wants to do .... Their own technology strategies say what projects they want to do .... When that's signed off on the appropriate level, they can then go off and do R&D. That's reported via a monthly reporting and various processes .... We're very much project-orientated, that's the way we work.

Synergies between divisions (for example, working together, linking cross-functional type) may also link in with the individual divisional technology strategies as well. Each division will get authorization on a technical project at the appropriate level and then carry out their R&D. This process is reported on via a monthly reporting process and is very structured, which reflects the project-orientated operation.

7.15 Opportunities for Improvement

The organization currently stores its knowledge in various ways; for example, in e-Mail databases, G-drive, computer files and inside people's heads. Although the knowledge management system is only a fledgling system, there is a strong organizational culture of support between individuals. The culture supports free sharing of information, due to the equally strong project culture which demands that projects are done on time, in specification and in budget. In order to do this, there is the idea that if you can give information to assist others to ensure a successful project, they must be supported. There are no barriers to asking and receiving, with the exception of a lack in time. There is also a sense that it is an individual's job to help in any way they can, and that the project's success is the number one priority.

There is a tension between time and cost, but each division still retains an information-sharing culture as much as possible. The practicalities are such that from time to time, "stove piping" can occur. However, the culture is to share, and it really is the company's policy to do so. McGourty *et al.* (1996) state that

> Exemplary organizations do several things: they involve outside experts; they exchange technology across the organization; they link rewards to new product performance; they use cross-functional teams; and they maintain other formal support systems such as job rotation, career paths, training, and communication linkages.

To enhance the organization's innovation capability, there have been strategic alliances, exchanges of information and active relationship building with external research organizations, such as DSTO, CSIRO and Australian universities and organizations. These are usually managed with contractual agreements. The company has also acquired whole companies, which includes the technology, skill bases with working teams and the licensing of software, tools and designs.

The R&D Manager stated,

> DPEC has developed distinctive core competencies, in that it tries to guarantee that all the trained and certified staff can ensure that products are fit for purpose and are safe to use. This could also be viewed as a contribution to greater competitive advantage.

Whilst DPEC carries out exemplary organizational structural practices which support innovation capabilities (and innovative products and services), these are done predominantly within the project and divisional structures. McGourty *et al.* (1996) discuss key strategies that exemplary organizations may undertake to support the necessary culture and behaviors that are conducive to innovative capabilities. Some of these are: monitoring outside events, recognizing critical technology for growth, focusing on core technologies, involving the customer and seeking out appropriate applied researchers for development. A reflection of DPEC's innovative capabilities can be linked to project needs and organizational structures.

Organizational strategies need to reflect structure (and other relevant relationships), which allows for both commercial and technical development and innovation. Ultimately, strategy can affect an organization's ability to innovate. Cooper (1985) discusses strategy as being significant to product innovation. Over time, DPEC has developed distinctive core competencies within the confines of the division and project structure. More recently, these capabilities have expanded into more horizontal aspects of the organizational structure; for example, with regards to successful overseas projects, the ability to offer that the project be carried out locally. In effect, localizing the product allowed for its ultimate success and for other advantages.

These key advantages were economic (and all that it implies) and process innovation. This process innovation occurred through prior project experience, expertise and relationships between Australia and its overseas client. The benefits from the previous projects have provided continuous improvements with regards to process innovation.

7.16 Conclusion

DPEC is very successful and experienced in designing, developing and delivering complex projects, products and services, using the combination of specific project know-how. Within the project framework, the organization must deliver complex products and services. This is where e-Communication (internet and intranet) has been used and has proven very beneficial to DPEC's success. Improvements of real time access to data could be supported by more development of data, information and knowledge management (KM) strategies, for example, access to lessons learnt through a more horizontal structure. This would also contribute to the reduction of "stove piping" for the small project teams.

Innovation strategies need to be supported by KM strategies so as to maximize the use of company experiences, both collectively and individually. The company is currently working towards strategies that are necessary for the development of horizontal exchange (expanding on the existing exchanges that occur with managers) with many more levels (people, groups), so that knowledge which is embedded in products would not be lost (usually occurs due to being primarily project-focused).

The strength and dynamics of SDTs in relation to the three knowledge domains — new product development, e-Communication and sustainable development — contributed to the development of DPEC's innovation capability phases, assisted with the innovation process. The case study also found that innovation capability, which already existed in the organization, had linkages to the three levers. Based on these findings, we conclude that innovation capability and innovation levers can have correlating dependencies with each other during the innovation development.

Review Questions

(1) How is innovation capability developed and knowledge transferred in the product innovation process at DPEC?
(2) Explain how DPEC manages and exploits its knowledge.

Chapter 8

Drivers of Innovation Capability for Effective Sustainable Development: Best Practice at Vaisala

Milé Terziovski and B. Sebastian Reiche

8.1 Introduction

The field of innovation management is increasingly viewed from a resource-based (Wernerfelt, 1984; Prahalad & Hamel, 1990; Barney, 1991) and dynamic capabilities (Teece & Pisano, 1994; Galunic & Rodan, 1998; Eisenhardt & Martin, 2000) perspective that focuses on how a firm can constantly exploit and reconfigure its idiosyncratic resources in order to create innovative output and thus maintain competitive advantage over time. In this context, the literature conceptualizes innovation capabilities as the key drivers of effective innovation (Lawson & Samson, 2001) and discusses various elements of these organizational capabilities, such as innovation strategy (Kim & Mauborgne, 1999), a supporting organizational structure (Burgelman & Maidique, 1988) and heterogeneity of knowledge (Rodan & Galunic, 2004). Recently, additional factors like sustainable development (SD), e-Commerce and new product development (NPD) have been discussed in terms of their potential to strengthen a firm's innovation capabilities and ultimately improve its innovative output (Metz *et al.*, 2004).

In light of recurring natural disasters, the sector of environmental measurement has been subject to increased public and academic interest, supported by progress in related measurement technologies (Morss & Hooke, 2005). For example, empirical evidence suggests that more accurate hurricane forecast information has a substantial financial impact on different industries (Considine *et al.*, 2004). This development is likely to continue to increase innovation pressures in this industry. In addition, as environmental management issues are an inherent component of business models in this sector, the industry serves as a valuable field to study the particular role of SD as a driver for innovation capabilities and its effect on other elements of innovation capabilities. Adopting a qualitative case study design (Eisenhardt, 1989; Yin, 2003), this study investigates the role of SD, e-Commerce and NPD as key elements of firms' innovation capabilities in the environmental measurement sector, drawing upon evidence from Finland-based Vaisala.

The case study is part of a larger research project funded by the Australian Research Council, whose aim it is to compare Australian companies with international "best" innovation practices, thereby detecting competitive gaps for further improvement in innovation. The case study is based on an in-depth interview with the Managing Director of Vaisala's Australian subsidiary based in Melbourne. To neutralize possible mistakes or misunderstandings inherent in a single research method, we conducted data triangulation and supplemented the interview data with a series of company information like company brochures, the company website and annual reports (Miles & Huberman, 1994; Yin, 2003).

The text comprises four sections. First, we characterize the case company with regard to its corporate background, core competencies, mission statement, resource availability and innovation strategy. Second, we draw upon a conceptual model of innovation capabilities that serves as a framework to analyze our data. Specifically, we view innovation capabilities as the result of an integration of various influencing factors, placing a specific focus on SD, e-Commerce and NPD. We then examine the relevance of these three potential drivers of innovation capabilities for Vaisala. The text concludes by highlighting organizational and research implications and summarizing its key findings.

8.2 Company Background

Vaisala, a Finland-based multinational company in the environmental measurement industry, is a global market leader of meteorological equipment and related services. The parent company is headquartered in Vantaa, Finland and has a workforce size of over 1,000 employees worldwide. Vaisala maintains offices and business operations in the United States, Canada, the United Kingdom, Sweden, France, Germany, China, Malaysia, Japan and Australia. Vaisala offers a comprehensive range of products that provide the measurement data necessary for forecasting the weather, protecting the environment and improving the safety of air and road traffic. In industrial settings, Vaisala products help to enhance the efficiency of manufacturing processes and improve the working environment, as well as reduce adverse impacts on the environment.

The company comprises three divisions, each focusing on a particular area of environmental measurement, namely Vaisala Solutions, Vaisala Measurement Systems and Vaisala Instruments. Vaisala Solutions supplies systems, services and solutions for observing weather conditions on the earth's surface and for collecting, managing and storing surface weather data. Vaisala Measurement Systems develops, produces and markets instruments and systems for observing the weather in the upper atmosphere. The division's products and services are used in many weather measurement and observation applications, and include radiosondes for upper-air observations, ground equipment that receives and processes the weather data transmitted by the radiosondes, wind profilers and lightning detection and localization systems that make extensive use of remote sensing technology. Vaisala Instruments supplies instruments for the measurement of relative humidity, dewpoint, barometric pressure, carbon dioxide, oxygen, wind, visibility, cloud height and present weather. The division also offers maintenance partnerships for these instruments.

Vaisala products cater for the environmental measurement needs of customers operating in many different fields, with a major focus on the government sector. Vaisala's core customer groups are meteorological and hydrological institutes, defence forces, aviation organizations, road and rail organizations, integrators of meteorological systems, companies with weather-related needs and industrial companies. In 2006, the company generated net sales of 220.8 million euros, with operations outside of Finland accounting for 97 percent of net sales. Vaisala Measurement Systems contributed the highest share of sales (42 percent), followed by Vaisala Instruments (29 percent) and Vaisala Solutions (29 percent).

8.3 Core Competencies

Vaisala's operations are based on two main core competencies that the company continues to develop. First, Vaisala maintains a high knowledge and research intensity that translates into superior innovative outputs and has made the company lead technological progress in its immediate field of operational focus. For example, Vaisala has been a pioneer in the development of relative humidity sensors ever since 1973, when the world's first capacitive thin-film humidity sensor HUMICAP® was launched. Products based on HUMICAP® Sensors are now used in industrial building automation, meteorological and agriculture applications. A related aspect concerns the company's high priority of competence development with regard to its organizational processes, tools and skills of its personnel. Indeed, Vaisala boasts a high skill and competence base with 41 percent of its global workforce holding a university degree. The resulting heterogeneity of knowledge residing within the firm's employees (Grant, 1996) is thought to lead to superior innovation performance (Rodan & Galunic, 2004). Second, Vaisala distinguishes itself from its main competitors by offering a comprehensive range of applications and services with regards to environmental measurement and meteorology, thereby serving as a total solution provider for its customers:

> Our main strength is that we cover pretty much all applications in meteorology, whereas our competitors take out only a small slice of the business.

As a result, Vaisala is able to provide integrated solutions that can be tailored to customer needs. This approach provides the company with a substantial differentiation advantage.

8.4 Mission Statement

Vaisala's mission is to provide environmental measurements that create the basis for a better quality of life, protect life and property, optimize economic activities, promote environmental protection and improve the understanding of climate change. The company places a high focus on innovativeness and is driven by six core values, namely customer focus, science-based innovation, goal orientation, personal growth, focus on greater good and fair play. This mission hence reflects a corporate philosophy that builds upon a team-based, people-orientated and customer-focused approach to innovation.

8.5 Resource Availability

The Vaisala Group maintains high investments in its R&D infrastructure, supported by a strong solvency ratio that amounted to 81 percent in 2006. Overall, the company allocates 10 percent of its net sales to R&D, leading to a strong R&D position relative to its competitors. As the respondent highlighted, a main reason for the company's strong research focus lies in its ownership history:

> Despite the highs and lows throughout the company's lifecycle, our investments into R&D have never been touched, and that is probably because of the history of the company. Professor Väisälä, the firm's founder, had a strong research background and, for example, invented the radiosondes.

8.6 Innovation Strategy

The company's innovation strategy evolves around three pillars, namely innovation-based alignment of its organizational structure, customer orientation and internalization of knowledge from external sources. First, the company constantly aligns its organizational structure to follow the needs posed by its innovation and NPD focus. For example, Vaisala recently streamlined its division and product distribution structure in order to create better conditions for the continuous improvement of its business models. Within the scope of this streamlining process, the divisions restructured its business areas into the current four divisions mentioned earlier.

Importantly, the company maintains two different R&D departments and R&D directors in order to streamline its operations along different customer needs and innovative outputs. While one area concentrates on incremental research that enables continuous improvement of the firm's products and processes and refines the company's applications, a second department focuses on strategic research that intends to anticipate potential customer needs in the future, thus developing and extending the targeted market. While the former research activities are driven by the aim of satisfying existing customer needs, the latter are related to innovations that create new needs and demand.

This understanding relates to the current thinking of Kim and Mauborgne (1999, 2004), who illustrate the superiority of organizations pursuing what they call a value innovation strategy. This concept places a strong emphasis on the creation of new demand. They suggest that organizations need to reach beyond the fundamental extension of customer value in existing markets and focus on the development of potential customer needs into new markets, hence creating new customers.

Second, Vaisala maintains a strong customer orientation in order to tailor its innovative applications to its customers' present and future needs. In addition, Vaisala strives to provide comprehensive customer support, which is considered an essential part of the Vaisala service concept and encompasses maintenance, training and calibration services on a global scale. Existing research (e.g., Longenecker & Meade, 1995) confirms that local knowledge of the respective customer contexts and close customer relationships are main success factors for sustained customer service.

Third, Vaisala leverages its corporate skill base through a wide array of external collaborations with specialists, research institutions and joint venture partners. For example, the company sustains close cooperation with leading research institutes such as the NOAA (National Oceanic and Atmospheric Administration, USA), NCAR (National Center for Atmospheric Research, USA) and VTT (Technical Research Center of Finland). By complementing and extending the internal skill base and expertise, Vaisala is able to increase its absorptive capacity and the resulting knowledge intensity for the development of superior innovations (Cohen & Levinthal, 1990; Zahra & George, 2002).

8.7 Innovation Capability Model

Existing research has mainly concentrated on identifying different elements of innovation capabilities without examining possible interaction effects between these factors and differences in their relative contribution to a firm's innovation capability (Metz *et al.*, 2004). However, there is evidence that the relative importance of these factors differs. For example, Samson and Terziovski (1999) demonstrate that HRM factors seem to be more important than non-HRM factors to a company's capability of benefiting from ISO 9000, Total Quality Management and innovation initiatives. At the same time, with pressures for innovation growing on an international scale, new potential drivers for innovation capability have emerged such as sustainable development, e-Commerce and new product development (Dunphy *et al.*, 2003; Metz *et al.*, 2004). However, there is still a paucity of empirical research to determine their relevance and contribution to a firm's innovation capabilities.

In light of this research gap, an integrative model of innovation capabilities has been suggested to explore the role and interrelation of these three factors with regard to a firm's innovation capabilities (Metz *et al.*, 2004). Specifically, the researchers propose that expertise gained in these three fields enhances a firm's innovation capability, and that an integration of these three factors increases the size of this effect. Fig. 2.1 in Chapter 2 illustrates the proposed model. We now turn our attention to the role of the aforementioned factors as drivers of innovation capabilities at Vaisala.

8.8 Drivers of Innovation Capabilities at Vaisala

Sustainable Development (SD)

Vaisala incorporates environmental considerations into its business operations as a result of both external influences and internal drivers.

On one side, the company ensures compliance with environmental government regulations concerning an environmentally friendly conduct of business. To achieve this conformity on an on-going basis, the company has established an ISO 14000-based environmental management system that addresses and constantly monitors all key environmental indicators. Research confirms that the establishment of environmental management systems is a powerful tool for multinational companies to stay ahead of environmental regulations that differ widely across countries they operate in (Sharfman *et al.*, 2004). Other external influences to incorporate SD aspects stem from customers and suppliers in terms of ensuring the use of compatible processes and interfaces. At the same time, environmental protection and concern are an integral part of the company's mission and are deeply embedded in the corporate philosophy. In fact, as the respondent noted,

> Sustainable development as a strategy was not implemented due to external forces, but has always been a substantial part of our corporate philosophy.

- *Vaisala's environmental policy (adapted from Vaisala)*

(1) As a manufacturer of products for environmental measurements, we aim at continuously improving our products, services and processes as to their environmental aspects, prevention of pollution and reduction of waste.
(2) Vaisala's products and processes comply with relevant environmental legislation and regulations, and other criteria to which Vaisala has subscribed.
(3) In development of new products, our aim is to minimize environmental impacts by managing the product chain from design to disposal. Design for environment is one of our basic design rules.
(4) Specific environmental objectives and targets are defined in division and strategic business unit (SBU) strategies.
(5) We measure, review and improve our performance with the help of environmentally-oriented key indicators.

(6) Vaisala's environmental management system is based on the requirements of EN ISO 14001 Standard.

It becomes clear that Vaisala has not only embedded external regulations into its basic operating procedures, but systematically attempts to build its own competencies in commercializing applications in an industry that is closely intertwined with environmental considerations. For example, Vaisala's ice warning and prediction systems produce real-time weather and road condition data to support and schedule maintenance operations. As a result, the environmental load is reduced as the winter maintenance is optimized. An important characteristic of SD at Vaisala is the fact that the company has incorporated environmental responsibility over the whole product lifecycle. Vaisala intends to minimize environmental impacts right from the product development stage. In fact, environmentally-oriented product design is defined as one of Vaisala's basic product design rules. The objective of "design for environment" is to improve energy efficiency, reduce material usage and improve the recyclability of products. As the respondent stated,

> Sustainable development is a core part of our product development and these considerations are inherent in every stage of our products' lifecycles. When we design new products, the development takes place in the light of ensuring continuous environmental fit.

Consequently, SD has become a distinct business process that cuts across the firm's different functions. In this regard, SD considerations are substantially linked to new product development and point towards the existence of interaction effects between sustainability and new product development in terms of innovation capabilities.

Vaisala sees a main benefit from its integration of sustainability considerations into business operations in the building of trust and respect towards its customers and the wider public. This is in line with existing research that has identified competitiveness, image and legitimation as well as social responsibility as main drivers for corporate environmental responsiveness (Porter & van der Linde, 1995; Bansal & Roth, 2000).

At the same time, the respondent at Vaisala highlighted that availability of and access to adequate technology are key drivers in extending SD considerations within the business.

To summarize, SD at Vaisala forms not only part of an on-going aim to comply with external environmental legislation, but, more importantly, is deeply embedded in the corporate philosophy and translates into a key competence and enabling factor for the firm's innovation capability.

e-Commerce

Vaisala's main applications of internet-based technology in its innovation process entail a company-wide intranet and e-Procurement on a global scale. While the company has established an integration of suppliers into the corporate intranet via extranet applications, the corresponding integration of customers is still in the development stage. The adoption of internet-based technology at Vaisala has been primarily driven by the requirements of knowledge storage and diffusion across different parts of the company. This is even more important as the company boasts many geographically dispersed units which makes knowledge exchange more complex. As the respondent noted,

> Access to company-wide information is particularly important for remote offices where access to traditional sources of information can sometimes take a long time.

Existing literature not only supports the notion of ease of knowledge transfer across internal organizational boundaries though intranet-based communication tools (Fulk & DeSanctis, 1995), but also emphasizes benefits of internet-based technology for the development of global strategy (Yip & Dempster, 2005). Importantly, electronic information exchange at Vaisala enables joint product development through the use of virtual teams across organizational and geographical units, thus pointing to another benefit of internet-based technology (Salazar *et al.*, 2003). This finding leads us to conclude that there are also substantial interaction effects between the use of e-Commerce and new product development, thereby enhancing a firm's innovation capability. In addition, Vaisala adopts a highly decentralized approach to knowledge diffusion. Indeed, the process of lodging and accessing information over the intranet is dyadic and interactive. In this regard, the respondent stressed that

> The intranet can be accessed by everyone at any time and to the fullest extent. Everybody can participate, and I think over a hundred documents are added every day.

Through this approach, the company tries to nurture a knowledge-sharing culture that encourages its employees to actively diffuse individual knowledge and thus create and extend an organizationally valuable knowledge base. Finally, the respondent emphasized that internet-based applications increase efficiency and allocate valuable resources to the innovation process.

New Product Development (NPD)

New product development at Vaisala is primarily driven by customer demand, and therefore occurs in close cooperation with the customer. The respondent mentioned:

> We make all these products for our customers and therefore get them involved in the development process. So, in a sense, our customers have a great deal of ownership.

Accordingly, Vaisala has established a strong level of customer involvement which, however, is only formalized through confidentiality agreements rather than commercial contracts. This ensures the necessary flexibility to, for example, step out of a specific venture if competing resources need to be reallocated. Empirical research confirms these findings by demonstrating that customer orientation in innovation projects has a positive effect on NPD success and that the impact increases with the degree of product innovativeness (Salomo *et al.*, 2003).

Despite the main focus on customer involvement, Vaisala attempts to integrate the whole process of NPD by including the supplier side and making use of cross-unit development teams that, as mentioned earlier, frequently collaborate via electronic communication channels. This integrative approach to NPD requires a harmonization of sustainability considerations between Vaisala and suppliers, and emphasizes the close linkage between both factors in terms of their effect on the firm's innovation capabilities. This integration of NPD has also resulted in an acceleration of production processes. As the interviewee noted,

> We have substantially reduced our cycle times because when we are now developing a product, we design modules that can be manufactured on a parallel basis and then outsourced.

In this regard, the company, for example, reduced the assembly time in the factory from 40 minutes to only a few minutes.

8.9 Integration of Innovation Capabilities

The case analysis not only demonstrates that there seems to be a distinct influence of SD, e-Commerce and NPD on organizational innovation capabilities, but that this effect can be enhanced through a systematic integration of these three factors. First, the results indicate that there is a potentially close linkage between sustainability considerations and NPD. If a company's SD policy forms part of its defining values and corporate philosophy as is the case at Vaisala, it becomes an integral part of the firm's business operations and thus cuts across different functions. Importantly, it will directly affect NPD and lead to an environmentally conscious design of new products, thus serving as a potential differentiating factor. Given positive effects of ecological responsiveness on economic performance (Klassen & McLaughlin, 1996; Derwall *et al.*, 2005), this linkage is an important driver for superior innovation and competitive advantage.

Second, there also appears to be a linkage between e-Commerce and NPD. Internet-based communication channels enable the use of research and product development teams whose members are geographically dispersed and thus are not able to physically cooperate. Here, information technology and its company-specific configuration can provide ubiquitous access to and diffusion of internal knowledge for employees across organizational and geographical boundaries, allowing joint product development to take place on a global scale.

8.10 Supporting Capabilities

The case study indicates that there might be additional drivers and capabilities required for a firm to capitalize upon the integration of the proposed three factors. A first issue refers to structural aspects. The example of Vaisala highlights that a firm needs to align its organizational structure to its innovation strategy in order to proactively develop its market position and create new future demand. However, this argument can be taken one step further. As electronic communication channels become increasingly powerful, companies also need to align their organizational structure with their internet-based infrastructure. Again, the Vaisala case confirms this notion as the company displays a flat structure conducive to a participative culture of knowledge-sharing that, in turn, allows the firm to maintain a highly decentralized system of intranet-based knowledge diffusion.

Second, the effective use of globally dispersed research and product development teams is contingent upon a highly skilled workforce. While research acknowledges the importance of access to heterogeneous knowledge sources for innovativeness (Rodan & Galunic, 2004), employees need to have the appropriate skills to essentially make use of this knowledge. Therefore, it is the responsibility of the HR function to ensure constant competence development through training, international job rotation and frequent exchange.

Finally, it is important for companies to maintain a high knowledge and research intensity in order to be able to effectively recombine existing knowledge, create marketable research outcomes and translate the resulting new applications into the firm's product portfolio. In doing so, the company ensures that sustainability and NPD are constantly linked to leading-edge technology and are able to provide a differentiating advantage that can continuously drive innovation. This also entails the need to access knowledge from external sources to complement the corporate skill base and maintain a diversified workforce to increase the organizational absorptive capacity, thereby allowing for a higher level of knowledge to be processed, reconfigured and leveraged.

8.11 Conclusion

Building upon an integrative model of innovation capabilities, this paper investigates the impact of sustainable development, e-Commerce and new product development and their combined effect on innovation capabilities at Vaisala. The case supports the conceptual arguments in the literature that all three factors seem to exert an influence on a firm's innovation capabilities, and that their integration is likely to increase the effect. Furthermore, additional capabilities such as an aligned organizational structure, a highly skilled workforce and a high research and knowledge intensity play an important role in supporting the use of these drivers.

Several implications for innovation practice and research can be derived from the analysis. First, the case analysis empirically supports the notion that there is a wider array of factors that can drive innovation capabilities and, thus, innovation. Accordingly, managers need to cross functional, geographical and business unit boundaries and identify additional sets of driving factors. Importantly, the study indicates that different factors can exert interaction effects that will not materialize if these organizational factors are treated in isolation or in different parts of the organization without maintaining direct exchange. Additionally, there is evidence that sustainability does exert an effect on innovation outcomes, especially when linked to product development. Consequently, a strategic configuration of sustainability considerations can bolster a firm's competitive position.

Our research findings also call for more empirical research that builds upon the conceptualization of innovation capabilities as a multi-dimensional construct. Accordingly, integrative research designs are necessary to capture the isolated effect of the different variables, their relative weight and simultaneous effects resulting from possible interaction between various factors. Moreover, there are other factors that need to be controlled for. For example, it seems likely that the relative contribution of innovation capability drivers changes with regard to industry, the organizational lifecycle, international expansion and even the product lifecycle. Having adopted a qualitative case study design with rich interview data, our research provides the necessary foundations to empirically investigate the proposed relationships in more detail.

Review Questions

(1) Describe Vaisala's environmental policy in the context of its mission statement. Discuss the role that SDO plays.
(2) How does Vaisala nurture a knowledge-sharing culture?

Chapter 9

Developing Innovation Capability Through Intellectual Property Strategy in the Australian Biotechnology Industry: Starpharma

Milé Terziovski and Amy Lai

9.1 Introduction

This case study highlights best practice in intellectual property strategy and the successful development of innovation capability and commercialization of innovation at Starpharma. A case study protocol was developed and used in a face-to-face interview with the company's Intellectual Property Manager. Starpharma is an Australian biotechnology company whose core business is to develop and commercialize new pharmaceutical drugs based on innovation in the emerging field of dendrimer science. The case study explores Starpharma's strategic view and intellectual property practices by highlighting the strengths and challenges faced by the company in this area, in its pursuit of innovation capability to achieve commercial success.

Starpharma has a broad view of intellectual property that encompasses codified and tacit knowledge as well as people and relationships, which all simultaneously contribute to the development of innovation capability. This view highlights Starpharma's leveled business orientation. The company has the view that without human and fiscal resources to commercialize intellectual property, there would be little value to the organization. The company believes that it is important for its corporate culture to be diffused with values based on intellectual property. Its influence over all functions and roles adds a valuable layer of protection over the company's intellectual assets.

9.2 Innovation Capability and Commercialization Success

When a firm releases a new innovation into the public domain, there is a risk that the efforts involved in developing the innovation into a state of readiness for its intended use will then be undesirably imitated and reproduced. In such situations, hard-earned competitive advantage is severely diminished in an amount of time shorter than it was taken to accumulate. This kind of risk significantly reduces the incentives for innovation, when the innovator cannot be ensured some form of protection and "first rights" of taking an idea to market. Thus, innovation is composed of intellectual property.

Intellectual property is one of the outputs of an organization's business activities, or sourced through external acquisition or licensing. In business terms, intellectual property is commonly regarded as being linked to a firm's competitive advantage. Corporations that are able to sustain competitive advantages and enjoy commercial success know that there is a necessary mix of legal and non-legal mechanisms required to minimize the risk of losing ownership and control over their creations and commercialization opportunities.

9.3 Biotechnology in Australia

Advancements made in the broad field of biotechnology are increasing at vastly rapid rates. As an output from this trend, intellectual property represents significant opportunities for commercial exploitation. Successful organizations can be characterized as those that are attuned to these environmental sensitivities, understanding them well enough and able to adapt and manage internal means within their control when called for. The science of dendrimers is relatively new, discovered in 1979 by a team led by Donald Tomalia at Dow Chemical Company (USA). It was not until 1992 that dendrimers were available commercially, produced by Dendritech,

Inc. — a company founded by the same dendrimer pioneer, Donald Tomalia.

Today, Donald Tomalia has partnered with Starpharma to jointly establish the company Dendritic Nanotechnologies Limited (DNT), managing 33 key patent families, involving over 182 granted patents worldwide in the dendrimer science field. In the field of pharmaceutical drug development, the growing momentum and advancements in research and development of dendrimer technology are offering increasingly optimistic possibilities for global commercial applications.

In the last ten years alone, the number of publications and patents in the global field has risen tenfold to more than 1,500 in 2001. In particular, the pharmaceutical industry has been stirred by the possibility that applications of dendrimer-based nanodrugs will be capable of filling a market void where the adequacy of traditional "small" drugs has been deficient. Protection of intellectual property is at the core of the business for biotechnology firms. According to a recent study, intellectual property protection is the second most important external factor influencing companies' decisions to invest and use biotechnology. Companies were asked about the extent to which they made use of patenting as a means of protection for invention.

The main reasons to apply for patent protection are the safeguarding of the developed technology and its commercialization, the competitive advantage and the provision of better negotiation positions for licensing agreements. Participants in the study also confirm that the patent system is important for biotechnology entities as an incentive system for R&D investments. The degree of patenting depends very much on the degree of competition in the market.

Large firms, more than SMEs, confirm that their patents are being used as a tool to defend their technology. Biotechnology SMEs, which are more restrained by their economic resources, also use strategic patenting to achieve competitive advantages without expending too many of their own resources. The first prerequisite of strategic patenting is the active observation of competitors' patenting portfolios which is necessary to identify market niches and to place products in the right position in the market.

9.4 Background to Starpharma

The Starpharma Group is a R&D company that was established in 1996 to commercialize novel polyvalent compound technology discovered at the Biomolecular Research Institute (BRI), Melbourne, Australia. The BRI was formally part of the Commonwealth Scientific and Industrial Research Organization (CSIRO). Starpharma is one of the early pioneers in the field of dendrimer science. Dendrimers have shown potential in preventing the spread and growth of secondary cancer (metastasis). Starpharma is developing drugs, called "angiogenesis inhibitors", which act by reducing the growth of new blood vessels to growing tumours, therefore restricting tumour growth. Dendrimers have shown therapeutic potential in animal models of breast and colon metastasis.

Further angiogenesis inhibition may also have applications as a treatment for arthritis, retinitis and asthma. Angiogenesis inhibition is a new area of cancer research and there are currently no commercially available drugs in this area. The company's establishment was motivated by the vision that there is significant commercial potential for real-world applications based on dendrimer technology. Starpharma continues to be a global player, committed to transforming dendrimer technology into commercial applications for the pharmaceutical market.

9.5 Corporate Structure and Business Strategy

The Starpharma Group is an Australian-owned R&D organization consisting of three subsidiary companies operating under a parent funding entity established as a Pooled Development Fund (PDF). Starpharma Pooled Development Limited is a registered PDF and the primary funding entity for Starpharma. Since its inception, Starpharma has evolved its structure to reflect the activities it will concentrate on as it develops its intellectual property and licensing opportunities. There are currently three wholly-owned subsidiary companies in the Starpharma Group, namely:

(1) Starpharma Limited, which holds the license for the technology and manages the group's intellectual property portfolio. Starpharma

Limited also manages research programs, START grants and administrative functions for the group.

(2) Viralstar Limited, which has agreed to provide research funding for further development of the technology in anti-viral therapeutics. Viralstar may become involved in the commercialization process of this area of the technology.

(3) Angiostar Limited, which has agreed to provide research funding for further development of the technology in the field of inhibition of angiogenesis. Angiostar may become involved in the commercialization process of this area of the technology.

Private investors who had a vision for the commercial potential of the technology initially funded Starpharma. The company raised further capital through a public float in September 2000, when the group was listed on the Australian Stock Exchange as Starpharma Pooled Development Limited (ASX: SPL). Today, the group comprises three wholly-owned subsidiary companies: Starpharma Limited, Viralstar Limited and Angiostar Limited, which own 49.99 percent of DNT (incorporated in Delaware, USA). Starpharma's CEO's presentation to the 2002 AGM clearly highlights the foundations of its value proposition:

(1) Dendrimers are a basic building block of nanotechnology.

(2) Nanotechnology is the most exciting new technology of the 21st century.

(3) Nanotechnology has the potential to transform industries by producing a whole range of new products.

The company's 2002 Annual Report further reinforces its vision to exploit its strong technological and commercial competencies by:

(1) Developing high-value dendrimer nanodrugs to address unmet market needs;

(2) Enabling the incorporation of dendrimer technology into new opportunities; and

(3) Enabling the use of dendrimer technology to enhance existing drug products.

The translation of Starpharma's corporate strategy into the company's business strategy is equally well-focused. Starpharma has a vision for developing and sustaining its core competency as a technological leader of an emerging base technology — dendrimer science, and adding value through identifying and developing applications with commercial potential.

Coupled with these objectives is a major emphasis on defining and establishing relationships with other groups who can provide the necessary complementary assets required to realize commercial success in delivering dendrimer-based pharmaceuticals to a worldwide market. Starpharma Limited is the group's operational entity concerned with the management of research programs, grants and administration. Starpharma Limited is also the license holder of and manages the group's intellectual property portfolio. Viralstar Limited and Angiostar Limited are R&D entities specializing in the respective fields of anti-viral therapeutics and the inhibition of angiogenesis.

9.6 Workforce and Culture

The Starpharma Group is a "lean" organization comprising 22 employees, more than half of whom are engaged in R&D capacities. The workforce is highly educated, with 11 employees holding PhD qualifications. The senior management team, representing one-third of the organization, can be characterized as possessing a multi-disciplinary outlook and strong appreciation of technology, with almost all members holding at least one qualification in the field of science.

The small size of the workforce is conducive to fostering heightened awareness in all employees of the business and activities in which Starpharma is involved. Roles and responsibilities for individuals within the organization are broad, and not limited to specific technical functions; this is a product of the culture at Starpharma. An example of the culture can be observed by the response of the Intellectual Property and Commercialization Manager (IPCM) upon discussing formal procedures for managing intellectual property. He indicated that:

Intellectual property is embedded in our culture.

DNT is an entity that operates autonomously from Starpharma, with an independent organizational structure of 15 employees. Although in the initial stages of the investment relationship, Starpharma was an active collaborator, supporting the maturation of DNT's capabilities in the areas of operational effectiveness, as well as providing input into the direction for R&D activities.

9.7 Core Activities, Products and Services

Starpharma is dedicated to the research and development of new chemically engineered drugs. The company focuses its pharmaceutical development efforts around the dendrimer platform technology (class of compounds). Its development objectives are differentiated from other industry players in that the company's dendrimer patents are based around the use of dendrimers as pharmaceuticals, rather than use as drug carrier molecules. At the core of this differentiation is the high value-added and significantly "complete" nature of its pharmaceutical products, thus positioning Starpharma in a strategically different relationship model with "Big Pharma" than is predominant with many other biotech companies.

One product example is the ViaGel, one of Starpharma's most advanced products from the company's core development focus area of the prevention of STDs. The ViaGel is a topical vaginal microbicide gel that is currently being prepared for submission as an investigational new drug (IND) to the Food and Drugs Administration (FDA) in the United States. The cutting-edge nature of this area of drug development will see Starpharma as the first applicant to undertake human trials of a dendrimer nanodrug. Starpharma accelerates the lengthy new product development (NPD) cycle of pharmaceutical development by marketing two early generation dendrimer products available through the international chemical supplier, Sigma-Aldrich.

These product offerings aim to establish an awareness of Starpharma's capability to produce quality materials. In addition to a growing product range, Starpharma offers specialized dendrimer-related analytical services (mass spec, NMR, CE, etc.). The offering of consulting and contract services draws attention to the industry's regard for Starpharma competency and expertise in the field of dendrimer science. The investment in DNT provides Starpharma with access to DNT's rich intellectual property portfolio, which currently consists of 33 patent families (182 issued patents), strengthening its position to consider new applications for nanodrug and accelerate the product development process. Currently, DNT markets approximately 12 dendrimer products through the Sigma-Aldrich catalogue.

9.8 Intellectual Property Strategy

Starpharma is a company that grew out of a need for a commercial vehicle to transform an existing set of patents into commercial reality. Since its establishment, Starpharma has developed and acquired, directly or through licensing, access to a broad portfolio of intellectual property. The major focus is on product development. Starpharma views itself as a world leader, taking the first dendrimer-based drug through the regulatory process for testing in humans.

From a definition point of view, Starpharma chooses to define its intellectual property in the broadest sense to include its human capital, intellectual assets and intellectual property (as legally defined). The company's human capital incorporates key relationships, skills, creativity, institutional memory and know-how. Intellectual assets include drawings, programs, data, inventions and processes. Legal regimes used to explicitly contain intellectual property include patents, trademarks, trade secrets and copyright.

9.9 Alignment with Business Strategy

Starpharma believes that the development of new pharmaceutical entities relies heavily on the effective and efficient creation, protection and commercialization of intellectual property in the global marketplace. The company considers its intellectual property to be fully integrated with its business strategy, as emphasized by the IPCM:

> Matters related to intellectual property are a constant theme in the company's strategic planning and product development processes.

Starpharma has an adaptable business strategy that continuously evaluates corporate intent with the evolving nature of the external market. This adaptability is evident in Starpharma's intellectual property strategy. The company has successfully aligned innovation and activities with current business objectives. With reference to business risk, Starpharma indicates that the decision, if and when, to seek legal intellectual property protection, is constantly weighed against the costs and benefits of the protection regime. For example, if the risk is in terms of losing competitive advantage, then the benefit to establish formal claims for protection can be justified. In spite of holding a view that intellectual property (in a legal sense of patents, trademarks, etc.) is a major factor for corporate success in alignment with the company's broader definition of intellectual property, Starpharma expresses its most valuable asset as being

> ... the relationship it maintains with its dedicated employees, followed by its cash resources that allow the company to maintain its R&D activities. Both of these activities are more important to the company than its [legal claims to] intellectual property, recognizing that Starpharma's intellectual property is essential to the future success of the company.

9.10   Protection and Management

Starpharma considers the possibility of being the first to successfully deliver a dendrimer nanodrug for human application to market as having greater sustaining benefits than merely owning a rich patent portfolio. In the current uncertain and immature state of the commercial market for dendrimer nanodrugs, Starpharma feels that minimizing lead time to market is a strategy that offers more effective protection over the company's intellectual property than patent protection. Implementation of this strategy is observable through the company's decision to make two early dendrimer products commercially available. According to the IPCM, Starpharma is proactive in its marketing activities:

> Get Starpharma's name out there to make companies aware that we have the capability to produce materials of that class.

The company views that the progress made to date towards achieving this goal has had significant impact in enhancing the company's reputation and displaying, to potential commercial "Big Pharma" partners, Starpharma's technological and operational competencies and capabilities. Starpharma has a structured approach to innovation, with formal corporate plans to pursue and commercialize innovation that are updated regularly. However, there is low formality in terms of codified procedures to identify patent-worthy discoveries. The IPCM stated the risk in having prescriptive procedures as follows:

> If you apply a hard and fast rule once, the second time you apply it, it's the wrong rule or you've applied it the wrong way, because there are so many elements to consider.

Rather, Starpharma's approach to disseminating practice concerning intellectual property is based on developing the corporate culture:

> In all of our meetings, IP [intellectual property] is a very, very central element to the conversation. It's really embedded in what we do all the time.

The stages which its current products and R&D projects are at ensure that the topic of innovation and commercialization are regular topics of discussion at senior management meetings. Management plays a key role in creating and maintaining a corporate culture that is conscious of the importance and sensitivities of intellectual property. Structured management of intellectual property in terms of patents and other forms of legal protection is centralized.

Starpharma Limited is responsible for coordinating with the relevant management, technical and legal parties in the drafting and filing of patents and trademarks. Starpharma chooses to make initial filings in Australia, and subsequently other major markets, including USA, Europe, China, Mexico, Canada and New Zealand. The company has a strategy to take out broad patent claims, and where possible, to block other players in the field from making challenges into their area of core intellectual competence. A number of dormant patents are maintained in the portfolio for this specific strategic purpose. Regular review of the patent portfolio in conjunction with the current business strategy influences decisions to develop new patents around older patents nearing end-of-life, or to allow other dormant non-strategic patents to become lapsed.

At this stage, Starpharma believes that the cost-benefit of trademark and copyright protection of brands is not an effective way to protect competitive advantage and as such, is not a high priority. The company's current strategy is to pursue Common Law protection for trademarks rather than formal legal trademark registration. The rationale for this strategy is largely influenced by its targeted objective to secure licensing arrangements with "Big Pharma", whom they recognize have their own agendas related to marketing matters:

> There is no certainty at all that the licensee is interested with the name associated with the product, because their marketing group may not consider that the name will meet their needs.

9.11 Networks and Collaborations

Starpharma has an enviable absorptive capacity in relation to leveraging technical and market expertise. The company is actively involved in acquiring knowledge and makes contributions to the growing field of dendrimer science. Starpharma's view on its involvement in the growth of the dendrimer field was described by the IPCM as:

> ... empowering competitive activity through partnerships, in a sensible umbrella arrangement, to work together to solve problems. We have this philosophy of lowering barriers to entry to allow and encourage other companies to invest in the development of dendrimers.

This is a considerably altruistic view, one that advocates cooperative effort rather than doing it alone. Executive and technical members alike are active participants, formally and informally, within scientific collaborative partnerships, alliances and the wider pharmaceutical community. Such networking opportunities are of significance to assist Starpharma in understanding emergent trends in the market and to be able to evaluate and direct its resources to the projects that offer best opportunities for success. Patent searching and working with specialized consultants are other ways in which Starpharma keeps actively informed of industry developments and maintains an awareness of the evolving nature of the competitive landscape.

Starpharma has a broad range of channels through which the company sources intellectual property. Internally, intellectual property generation is achieved through the continual creativity of the company's research scientists. Complementing internal capabilities, the company extensively leverages external channels including acquisition of intellectual property rights via both the in-licensing of intellectual property (via the BRI and DNT) and also from collaboration efforts through its global research network. The strategic investment in DNT, Inc., provides Starpharma with access to specialized technical competencies.

The structure of the relationship between DNT and Starpharma is such that Starpharma, a core partner for development and commercialization of pharmaceutical applications, can influence the direction of DNT's research and development activities. Starpharma's global research network, spanning widely across Australian, American and European research groups, affords the company similar influence to generate relevant intellectual property. Access to diverse sources of intellectual property serves to accelerate the investigative and product development processes. In addition to intellectual property sourcing, Starpharma's network of contract partners provides complementary capabilities in the areas of testing and manufacturing.

9.12 Resource Allocation

Subsequent to its public listing in 2000, Starpharma has taken significant steps in the allocation of resources to facilitate achieving its business objectives. The company became one of the first to locate a new laboratory and offices within the recently opened Baker Heart Research Building, part of the Alfred Medical Research and Education Precinct, Melbourne. The state-of-the-art laboratories and access to some of the best possible facilities and biomedical resources provides Starpharma with the necessary tools to perform best practice R&D. A significant proportion of the R&D activities undertaken at Starpharma are based on technology developed by others, of which Starpharma has acquired exclusive patent licenses. For the 12-month period ended 30 June 2002, Starpharma's total R&D expenses were approximately AUS$6.2 million. The in-house incremental approach to innovation of these technologies is complemented by efforts to invite and integrate skills and expertise from "best-in-class" scientists, whose motivation is largely seen in being able to participate in shaping the frontier of an emerging field of science.

From a strategic and operational perspective, considerable human resources have been devoted to organizational and managerial change, specifically in the areas of strategy planning, OH&S training, HR training, and HR restructuring. Two initiatives are notable: firstly, the establishment of in-house management capability for pre-clinical drug development. This is an initiative that aims to increase Starpharma's control over product development projects, with the intention of reducing development time, cost and risk. Secondly, investment in DNT drew on Starpharma's resources to develop strategic alignment that would be beneficial to both Starpharma and DNT. Starpharma was also able to lend its business and commercialization expertise to DNT, and will continue to provide input into directing future R&D activities undertaken by DNT.

9.13 Innovation Capability

In the last three years, many new lines of products have been conceptualized and evaluated by the Starpharma Group. The dedicated intellectual property management entity Starpharma Limited has, as one of its main foci, to identify new products related to the commercial application of dendrimers as pharmaceuticals, and has, for example, identified numerous product opportunities in anti-viral, anti-cancer, anti-toxin and bioprotection applications. The "market-pull" demand mechanism influences Starpharma's approach to new product development opportunities and the company is active in identifying and effecting incremental changes in its product and service lines.

This rate of innovation is common in many of Starpharma's current initiatives, especially since products are still in their relative early development phases. For example, Starpharma's drug products like ViaGel are not yet marketed in the final consumer market, and thus, relatively minor changes are still being adapted for their eventual application. Currently, Starpharma is involved in seven R&D projects focused on a wide range of biomedical areas — sexually transmitted diseases, systematic anti-virals, biodefense, respiratory disease, oncology, tropical diseases and new dendrimer architectures. Members of the company actively engage in making contributions to the field through published works, participation in conferences and involvement in nationally and internationally coordinated efforts that further developments in this field.

The priority placed by countries like Australia and the United States on furthering the field of dendrimer science, coupled with Starpharma's recognized core technological expertise in the field, has increased the organization's ability to attract public funding. This in turn positions Starpharma as a "hub" to access the best talent without losing the IP generated from the collaborative development efforts. Starpharma's global network consists of 10 local and 14 international research groups. Engaging in R&D at this level involves activities that include validating other global work in this area, reverse engineering and coordinating multiple perspectives, which serve to reaffirm Starpharma's position as a major global player in this field.

9.14   Implementation of Protection and Management

The Starpharma patent portfolio, managed by Starpharma Limited, consists of eight families in the areas of anti-viral therapeutics and inhibition of angiogenesis. Currently, Starpharma has successfully applied for patents in five of eight patent families in the major markets considered to include Australia, New Zealand, USA and Singapore. Patenting is recognized to be an essential formal mechanism for Starpharma to establish protection over intellectual property.

The ability to identify and evaluate patent opportunities is initially handled informally. The company does not have a formal audit process; regular reviews are conducted on an on-going basis to understand the intellectual property landscape. Starpharma involves external consulting partners in major markets, who are familiar with the field and the current intellectual property landscape. Starpharma's collaborations with external groups present a challenge to effective management of interactions in the areas of general scientific processes and efficiency, and particularly around issues of intellectual property rights and ownership. However, Starpharma has had positive experiences with academic collaborators and national research groups:

> Long-standing practice of commercial entities owning outright the intellectual property, the National Institute of Health (NIH) [in the United States] and other groups are really just interested in rights to publish, under appropriate delay arrangements if required.

The IPCM indicated that there is an obligation attached to the funding sourced from the US government, which maintains a right over intellectual property generated from research funded by it:

> If the government in the US is sufficiently excited by it [research funded by the US government], then, in theory, they could effectively force us to take a compulsory license. Their philosophy is very different to Australia. They have this idea that they will fund companies to carry out research of direct relevance to the government's priorities, and if there are any consequential commercial opportunities that flow over, fantastic.

The US government has the right to force any organization receiving its funding to continue with commercialization developments, even when the outcome is not aligned with the company's business priorities. The field of dendrimer chemistry is still a relatively new and specialized field, so there are still relatively few commercial players and even fewer focusing on the applications in pharmaceutical-related development. These current environmental conditions have not made it possible for Starpharma to determine the degree of effectiveness of patent protection.

Starpharma currently enjoys this competitive advantage, but also recognizes that this situation can turn against itself. This is why there is greater emphasis on protecting its position and intellectual property by minimizing lead time to market, combined with an emphasis on the management of secrecy and know-how. Starpharma's current view on trademarks, brand name and product marketing as a form of intellectual property protection is of lower priority and less important fit with their current business strategy, as their targeted objective is to secure licensing arrangements with "Big Pharma", whom they recognize have their own agendas for marketing matters.

Additionally, the nature of the current competitive environment is a factor in the company's current view, as expressed by the IPCM:

> ... particularly the microbicide area, a very new product concept —
> if it was a very crowded space, then we would be more concerned
> about it.

The company considers the maintenance of control over the distribution of key data and information to be of greater importance to protect competitive advantages.

9.15 Systems and Information Technology

Formal systems and information technology are enablers in daily operations at Starpharma. As a company whose activities are based around pharmaceutical product development, strict practices and compliance with regulatory agencies are required. Starpharma's implementation of a quality management system has been specifically developed in compliance with international standards and regulations, including those defined by the United States FDA, as well as the Australian Therapeutic Goods Association's (TGA) Codes for Good Manufacturing Practice (GMP) and Good Clinical Practice (GCP). In relation to contract and supplier contributions to Starpharma's product development efforts, the company's quality program extends to assure the quality of input and external service provided to the company.

Use of information technology is a prominent tool used daily within the company. R&D activities use information technologies for data collection and analysis. The bulk of data generated by Starpharma is the output of rigorous testing at each stage of the pharmaceutical product development process. Implemented quality systems also make use of information technologies of which the assurance of data integrity is a high priority. Starpharma believes that data accuracy and integrity is critical to reducing product development time. Staying connected with a global community and with what competitors abroad are doing can be achieved conveniently through access via the internet. Activities such as patent searching, accessing updated scientific and competitor information are now readily accessible through updated online sources.

Similarly, the dispersed nature of the organization's research network relies on cost-effective internet and e-Mail-based methods to interact and communicate designs, results and data. Internally, Starpharma's network provides multi-user access to knowledge directories and manuals for procedures and processes. Investigation for information technology solutions for knowledge management and collaborative project management has been considered to enhance current practices.

9.16 Conclusion

Starpharma is a specialized biotechnology company focused on developing a base technology around dendrimers with broad pharmaceutical applications. Unlike major pharmaceutical companies with all the necessary complementary assets to make drug commercialization realizable, the Starpharma business strategy recognizes the need to partner with the right individuals and groups that can offer access to the necessary complementary assets required for successful commercialization. The organizational structure of Starpharma reflects a core focus on intellectual property and relationship management; central to facilitating this is the corporate subsidiary, Starpharma Limited.

A key strength of the company is its operation as a virtual organization — a small permanent workforce that is complemented by an extensive network of collaborators at both the supply and marketing ends of the product value chain. The transient nature of some of the relationships, especially collaborations with specialized scientific researchers, allows Starpharma to attract and work with some of the best-of-class individuals and groups in the field. Starpharma aligns its core technology focus on dendrimers science with national research priorities. This allows Starpharma to supplement its R&D funding with public grants from Australia and the US, and to still maintain control over generated intellectual property.

Starpharma recognizes that legal claims over intellectual property are fundamental. The company has an understanding that is consistent with its sector. Successful application for patent protection in the larger R&D and product markets like the United States, which has internationally recognized regulatory requirements, generally affords accelerated applications in other geographies. These mechanisms are essential factors in the product development process to accelerate the process from concept to market.

9.17 Implications for Managers

Intellectual property is a key factor of competitive advantage for biotechnology companies. An important implication for managers that has emerged from the case study is that protection of intellectual property alone through legal regimes like patents cannot lead to successful commercialization. Investment into legal claims over intellectual property needs to be considered together with business strategy and business risk. Other methods of protecting intellectual property like secrecy and management of know-how represent effective forms of protection that must not be overlooked. Equally deserving of recognition are management and capital commitment. It is important that all businesses recognize that they need an IP strategy to commercialize their products successfully and avoid pitfalls.

Some of the key elements in developing an intellectual property strategy are: ensure that the firm's ideas are new and avoid infringing the rights of others by searching the patent and trademark databases; instill a "first-to-market" philosophy in the corporate culture; develop an infringement strategy; educate the staff so that confidentiality is an espoused value; establish a link between the intellectual property system and the company's business strategy. Furthermore, trademarks should underpin the core of the company's brand and image building strategy. IP assets must be identified and valued to ensure that they are itemized in the business plan.

Review Questions

(1) Articulate Starpharma's key innovation strengths and weaknesses. Why has it focused on IP? What role did IP strategy play?

(2) Explain how innovation capability was developed at Starpharma. Make recommendations to management on how Starpharma could make further improvements to innovation capability.

Chapter 10

Development of Innovation Capability at Invincible Company in Thailand

Suthida Jamsai, Susu Nousala and Milé Terziovski

10.1 Introduction

The aim of this chapter is to investigate the area of management of innovation capability, a process that has not as yet been extensively researched in Thailand. It has the potential to provide in-depth solutions to managers searching for appropriate methods to boost the development of innovative products and ensure long-term success for their companies. The development and management of innovation capability involves an understanding of the emergence of innovative ideas, the evolution of practices in managing innovation capability and the replacement of traditional operating systems with more supportive processes.

Thai managers encounter difficulties in replacing traditional operating systems as they must operate in an environment where "copycat" industries exist, constantly weakening their brand image. The companies' products struggle to gain any advantages in either global or local markets. The growing problem for Thai managers is to overcome this replication by establishing strategies to quickly introduce new products to sustain their innovation capability and ensure long-term success. Managers also need to consider examples from other countries that have faced similar situations.

To date, research on innovation in Thailand has been developed from a background of the national potential for innovative performance. For example, Intarakumnerd *et al.* (2002) considered that a national innovation system results from strong links between firms, universities and government organizations in Thailand. Numprasertchai and Igel (2004) provide a similar view that new knowledge required for the country's industrial innovation relies heavily on diffusion by particular groups, such as universities conducting research projects in science and technology.

However, an interesting paper by Brooker Group (2001) poses the question whether there is an understanding of the recognition of innovative activities in Thailand. The paper goes on to cite examples of innovation in Thailand, such as technological innovative activities of industrial enterprises, including environmental and strategic innovation. Although this research offers information about initiation and improvement of innovation systems through various uses of new technologies, no attempt is made to describe the development of these activities or innovation capability being carried out within individual companies in Thailand.

Individual companies (in particular, SMEs) have very different motivations and experiences which influence their pathways to innovation capability. As Culkin and Smith (2000, p. 154) show through their research,

> ... leading edge qualitative research has now burst the myth that small businesses are simply scaled down versions of large enterprises.

Smaller companies do not have the scalability of larger firms, including the higher volumes for manufacturing. In this regard, SMEs cannot so easily follow innovation capability pathways supported by agile manufacturing and mass customization of the larger firms (Brown and Bessant, 2003; Bessant et al., 2001). The use of ICT, primarily the internet and e-Mail, have improved business competitiveness and opportunity for SMEs, allowing them to compete on equal terms with larger organizations (Chapman et al., 2000).

Haynes et al. (1998) discuss the reduction of cost for businesses to engage in exchanging visually complex or graphical information accurately via the internet. The internet not only reduces costs for accurate information exchange, but can do so globally, contributing to the innovation capability being carried out within individual companies in Thailand (Haynes et al., 1998).

This study sets out to investigate the management and growth of innovation capability and its role in new product development at Invincible Company Ltd in Bangkok, Thailand. The company's innovation processes are analyzed from product conceptualization to production using the Integrative Innovation Capability Model developed by Associate Professor Terziovski from the Centre for Global Innovation and Entrepreneurship at The University of Melbourne, Australia.

The model is used to develop company case studies with the aim of identifying and developing the processes of innovation capability using the three domains of new product development, e-Business and sustainable development. Analysis of the findings gained through the Invincible case study contributes to the clarification of the enhancement of innovation capability in companies in a number of countries, including Thailand.

10.2 Company Background

At present, Invincible Co Ltd designs and manufactures firefighting trucks from its Bangkok main office and three factories. At the main office, at which conceptualization of new product ideas takes place, most of the design and support is carried out by a staff of five engineers involved in product design; twelve technicians are involved in the product prototyping; and four administrators are involved in procurement and administrative tasks. In 1994, the company was involved as a sub-contractor in the design and manufacture of items of civil and marine defence, cable discs and telephone poles.

In 1998, the company recognized an opportunity to diversify into the rapidly growing information technology business and started to design marine logistic systems and other related information technology systems for the Thai navy. Today, the company focuses mainly on a single contract project involving the design and manufacturing of firefighting trucks for use in Thailand by the Department of Local Administration, distributed throughout regional offices around the country; and by the Airport Authority of Thailand, providing the latest high-tech firefighting trucks for the new Bangkok International Airport.

The company was enthusiastic about involving itself in the manufacture of innovative products. The interviewed manager stated,

> It is the challenge to grab a chance in developing new products or trying new processes, if we are sure that there will be customers out there waiting for better products. It doesn't mean that we can't find any customers for the products we were producing, but we rather let others sub-contract the work that's no longer attractive to us. I and my team easily get bored if we have to work on the same thing for years. I didn't jump into totally strange products, but I know that I could encourage my staff to draw on existing knowledge and our practical experiences of building a yacht, warship, cable disc, and logistic systems to translate into the new firefighting truck.

10.3 Company Strategy

The interviewed manager emphasized that to meet the customers' needs, the model of the new firefighting truck had to be designed, developed and built as quickly as possible. To achieve this, the company introduced a strategy of "we can do better" that involved regular staff meetings to exchange ideas and maintain awareness of developments. This improved communication allowed the manager to recognize the company's ability to compete for contract work with Thai authorities, offering the best quality and efficient products within budget and on time. The manager mentioned that:

> We are engaged in products that are needed solely by public organizations, not private. I knew that with the government budget, they could order the best firefighting trucks from any other country. We had to find the way to make them reconsider why they have to throw their money out of the country when a Thai company like us could offer the same functions and similar qualities. But, what we do better is we build it here and we could do it the way they like, to suit them perfectly for the country's conditions. We are closer to them and we could deliver the products as they required.

The above comments reflect the company's understanding of the importance of working closely with its customers to collaboratively develop new products. With this in mind, the manager created an environment in which the customers could get involved with the design team. A variety of activities were formulated to undertake and deal with customers' requirements. These are presented in Sections 10.6 and 10.7.

10.4 Current Level of Performance

The management believes that they are on track towards meeting their objective of developing 40 firefighting trucks within the first five years. At present, the company is working towards meeting the deadlines of two customers, the Department of Local Administration within the Ministry of Internal Affairs and the Airport Authority of Thailand. The manager is also preparing to bid for a Bangkok Metropolitan Authority contract that will involve the development and manufacture of similar firefighting trucks scheduled for use in all Bangkok districts.

10.5 Invincible Customers

Invincible is recognized by its current clients for its creative customization approach in meeting specific requirements and challenges. The manager revealed that:

> I pointed out the features and technical specifications of the trucks made in Europe, and how I could modify them and make them better. Actually, one customer had already imported the trucks from Europe and planned to build them by themselves, but did not have enough capability to do it. This customer chased me to see whether I was interested in building a better one. "Why not?", I told them. My team and I used to build warships that are far more complicated than firefighting trucks. I know the two products are not the same and I have to build up new rules to make the new product happen.

In modifying the European-made firefighting trucks, the staff at Invincible designed a new model to satisfy both the main customers' expectations and the country's conditions. The modifications included: enlarged water tank size and loading capacity; increased speed; increased movement and flexibility; more specific accessories and functions; and lower energy consumption. High levels of quality assurance were required throughout the many technical specifications, ensuring accuracy in the relevant operations by the various effective censoring devices. These quality assurance measures ensured that the company was able to present effective monitoring to its customers throughout the manufacturing process.

10.6 Product Innovation

The conceptualization and improvement of the design of the firefighting trucks were carried out at the main office and involved several company partners and customers. Staff with advanced technical knowledge provided accurate and detailed information for the required technical specifications, enabling the design team at Invincible to refine the conceptual product design.

Product design requirements originated from the customers needing to increase the efficiency of the firefighting trucks, but the final product concepts were heavily influenced by the Invincible manager and his staff. This process of collaboration of customers' clear requirements and the ability of the Invincible staff to bring these requirements to reality was achieved by adherence to the following management practices:

Organizational Structure

As part of its "we do better" strategy, the manager formulated many new activities and procedures that tended to be flexible and informal. These were apparent in the organizational structure, especially for the staff in the engineering design team where several new joint-teams, including staff from different units, were established. For the design team, engineers with a range of expertise, qualifications and specializations formed one large team. Team members were not representative of any particular unit, but acted in co-coordinator roles to provide diverse input for the new product designs. Generally, the team discussed and commented on improvements to outperform its competitors. The manager said that:

> The engineers were formally assigned responsibility for various functions since the last production, for which we were subcontracted for building a ready-designed warship. I sensed that the work I had organized might not work well with the new task because the previous one needed a strict schedule work program with a designated responsible person so as to meet the costs and dateline as specified in the contract. This time, when I have to develop the truck, I must apply different working structures, which is in a form that is more approachable and flexible.

The manager continued to explain how he assembled the team and encouraged them to work together:

> I pulled them together in one team, telling them to forget about the past practice of dealing with their own unit. I arranged a platform for group thinking, group discussion and team problem-solving. I also gave them freedom to come or not come to work when they could not think or they could not come up with new solutions. However, I had a rule that the team must always present and discuss in the meetings concerning the development of the product design. No excuse is accepted. At the early stage of the new product development, I wanted them to be there so they could comment on the things which could or could not function in the designed model.

This flexibility in organizational structure and work practices appeared to be a successful stimulus for the development of innovation capability. Interestingly, the manager reverted on occasions to more formal strategies of regular reviews of production processes, checks on compliance with customers' requirements and tight work schedules, but the flexible work structure was retained. This freedom allowed staff to consider customers' preferences, assess changes in the firefighting truck industry and make crucial decisions. The manager explained,

> Actually, how the new product came about definitely needed a different style of control. I cannot force them to think, think, and then

hurry up and give me a new idea. On the contrary, I am quite easy and flexible, happy to let them have time to think by themselves, knowing they will come back to me when they have something and want to share with me or the team. It is as simple as that. I will then get something out of them. However, they did not have freedom for that long. As you know, time is crucial and costly. If we took too long to get a new idea out, the customer would go somewhere else. I did have some arrangements to let them talk and share ideas. We do it in working hours, at night or weekends.

We have just the right size of team. It is not that difficult to call them in for brainstorming. When we got the clearer idea on the product concept, the design team knew what to do next. I then adjusted my role to be harder on them. I need to make sure every production line knows the production plans. If we do not have a timeline, we cannot serve the customer and we will be out of the business.

The above comments clearly indicate the successful use of a combination of two organizational structures — flexible but sometimes formal, according to the various requirements of different types of projects, products, development phases and/or production stages. The staff was able to respond to problems as they arose and the development of the new product moved from one phase to the next. In addition, the flat and lean structure of the Invincible organization was a crucial component in achieving an effective, flexible management style and an open structure that facilitated the development of an innovative product.

Rewards for Success in Product Innovation

Invincible had an established formal system of annual evaluation of staff performances for its 21 employees. This was a relatively simple program agreed to by the management board, by which employees who satisfied work targets received wage increases. Bonuses were also awarded to all employees at the end of each year as a result of the company's profits. However, this system was adjusted to recognize and reward innovative suggestions made by staff about the development process of the new firefighting trucks. The recognition and reward were not fixed or formal, but were determined by the manager, depending on the impacts of the ideas on the project. These impacts varied, as did the rewards — a gold necklace, an overseas tour package, a bottle of whisky, a gift voucher, money — showing an appreciation by the manager of the staff's efforts and commitment to "their" innovative project. The manager revealed that:

> I gave everyone a reward if that person could only think differently from the others. It could be just a small, tiny idea from the back of the room. It may be the funny idea or quite distinctive thoughts that may or may not link to the present product, but it may lead us to see things in different ways and make the product happen, or we could keep the idea to be used in the future. Who knows? I would rather make them realize that I am happy to pay back to them whatever ideas, good or strange ones. If they could only raise their thoughts to challenge or contrast to others, they will be paid with big or small incentives.

10.7 Product Innovation Process

To accelerate the production of a new firefighting truck onto the market, staff at Invincible participated in the various following phases:

Phase 1: Information Gathering

Information relating to firefighting trucks presently manufactured by and available from other companies was collected and analyzed. This was done mainly by the engineers' team that was also designated by the manager to oversee the product's conceptual design. This phase ensured that the team knew all the competitors' existing product details, specifications and advantages and disadvantages of firefighting trucks already on the market. Information was obtained from two main sources.

First, internet search — each engineer searched for details of all systems, assembly and components of firefighting trucks produced by other companies. The team spent two months collecting this data and a further one month comparing and understanding current technology in the production of a firefighting truck. Second, study visits to competitors' companies — the engineers whose major responsibilities were the design and construction of the new trucks joined the manager to visit several European companies selected as a result of information from the internet search. The visits focused on companies known for the production of high quality firefighting trucks for worldwide distribution. The interviewed manager provided an interesting reason for these visits:

> I wanted my engineers to see the real thing. By looking at pictures from the webpages, they would only see the "good side" of the product, not really knowing the weaknesses of the trucks that others have built. For their visits at each of the companies, they have to sharpen their eyes to look for both advantages and disadvantages of each of the trucks. I then ask them for their opinion on what they saw and which parts they could make better.

Phase 2: Innovative Knowledge-Sharing

The transition to phase 2 began when several conceptual designs had been completed and agreed to by the potential buyer. Three teams — design, production and procurement — were established to discuss the approach to each design and the various possibilities to produce a firefighting truck at Invincible. Members of the design team raised various issues and concerns regarding the design of each component or system. At the same time, members of the production team gave their opinion on appropriate materials for the proposed designs, made comments about useful information and feedback on the improvement of components and/or materials, and offered suggestions concerning the development of the functionality of the proposed product components. The procurement team highlighted difficulties regarding issues such as the availability of equipment, machines or raw materials from new and/or currently contracted suppliers to meet production schedules. The meetings of these teams were frequently arranged in formal or semi-formal environments every morning for approximately two months leading up to the finalization of the design or starting date for the prototype process. The manager's comments below show a detailed example of actual practices the teams faced during this phase of innovative knowledge-sharing:

> I chose to meet with the teams in the morning to talk about each particular part, element or component of the design to make a more efficient truck. For example, one of the engineers in the design team may start with a simple bolt, telling others why and how this bolt had to adjust to fit with the specific part for the connection of each of the joints. Other team members might suggest that the bolt would work better if the adjustment could be made to have more front surface and twist into the right position, not the left; or its color has to be green, not red, to remind users during safety operations, etc. A member of the procurement team may mention that this bolt type has a head for the left only, while the bottom part can be adjusted 360 degrees to fit with any joints; and it can be made available in various colors with suppliers in China with low raw material quality and cheap, but suppliers in Europe are no longer producing it; or that orders could be placed through American companies with longer delivery times, etc.

The meetings were recorded and copies were sent to all the participants to confirm their agreement on the design, production (prototype) process and procurement. These regular and detailed staff discussions on specific parts (such as a small bolt) were important in the saving of time during the innovation process. Also, the in-depth analysis of the usefulness and transference of innovative ideas into the production process assisted in the making of decisions regarding choice of materials, organizational and manufacturing strategies.

Phase 3: Transformation of Customers' Requirements into Product Improvement

The design team made adjustments as suggested by the other two teams and the adjustments were then presented to the customers for their consideration. The customers' latest requirements and recommendations were then incorporated into the design. However, they had to be done within the limitations raised by the production and procurement teams. The outcome of this stage was agreement from the customer on a final design and timeline.

Phase 4: Sustainability of the Product Development

Phase 4 started when the truck prototype was approaching the completion date and the production process was about to be launched. Monthly meetings were held during the final stages of prototype production to ensure compliance with the agreed design and to update any requirements due to customers' new demands and/or suppliers' limitations. This phase highlighted the problem of team members making alterations to meet new situations without informing other members. The aim of this phase was to give the team an opportunity to reconsider their previous practices and to overcome any issues they faced during each production process.

While it was possible to make alterations in the prototype trial stages, it was much more difficult to make them during full production. This led to the introduction of full comprehensive engineering schedules. Thus, similar practices as seen in phase 3 were "redone" or lessons learnt, to give the teams opportunities to gain and share important essential information and knowledge of the new truck. By sharing the lessons learnt, it provided the team members with alternatives in order to accelerate their decisions in reducing their production cycle times, whilst keeping efficient and maxim output for their processes and retaining the company's intellectual capital, which could be passed on to those who need it the most.

10.8 Conclusion

Invincible Co Ltd is an example of a company in Thailand where management has successfully introduced strategies and activities to manufacture an innovative product. The company developed extensive communication processes that supported many other new approaches, including development of effective procedures to track current information related to the new product development. This ultimately saved time in generating innovative ideas. The company developed innovation capability through strategies that involved understanding the emergence of innovative ideas and the evolution of specific practices in managing innovation capability.

Key success factors to innovation included approaches such as the "we do better" strategy, which replaced traditional processes with more supportive ones. Other important aspects undertaken included combining two styles of management — flexible (informal) and formal — to form a new strategy. This new strategy was reflected in the organizational structure, especially for the staff in the engineering design teams as several new joint-teams were established. These aspects and strategies could give other Thai companies insights into approaches for management that will help their products maintain and improve their local and/or global market positions.

Review Questions

(1) Discuss how Invincible Company accelerated the production of the new firefighting truck into the market. Synthesize the various phases of the product innovation process.
(2) Explain what is meant by the following statement: "This flexibility in organizational structure and work practices appeared to be a successful stimulus for the development of innovation capability."

Chapter 11

Multiple Cross-Case Analysis: Conclusions and Implications

Milé Terziovski

11.1 Introduction

The Integrated Innovation Capability Model developed in Chapter 2 was used as a framework to examine the influence of the drivers of innovation such as NPD, sustainable development, e-Commerce and their relationship with innovation capability. The research questions articulated in Chapter 1 are addressed in this chapter, namely:

What constitutes innovation capability in organizations, and how can it be developed and exploited?

What are the key drivers of innovative organizations?

Furthermore, the Lawson and Samson (2001) innovation capability construct discussed in Chapter 1 is used to identify innovation practices under the following dimensions:

- Vision and strategy;
- Harnessing the competence base;
- Leveraging information and organizational intelligence;
- Possessing a market and customer orientation;
- Creativity and idea management;
- Organizational structures and systems;
- Culture and climate;
- Management of technology.

Our aim is to explore each of the organizational innovation capability dimensions above to identify which of these, individually or in combination, are perceived by best practice innovative organizations as critical to effective innovation.

11.2 Vision and Strategy

Value innovation intimately links customer value with technology innovation (Kim and Mauborgne, 1999). Technology innovation on its own does not address buyer value, thus a new technology might not be accepted in the market as having value for the customer. Technology innovation tends to focus on solutions, whereas value innovation focuses on redefining the problem. While technology innovation can create an exclusive market for a new product, for example, via patent protection such as in the Starpharma case study, this avenue is not open to all organizations. For example, service organizations create value for their customers by integrating a range of items (products, service and environment). Thus, such organizations are open to competitors copying their approach. To minimize this "free-riding" and become a value innovator organization, they must ask two questions:

> Are we offering customers radically superior value? Is our price level accessible to mass buyers in our target market?

The Starpharma case supports the conceptual argument that a strategic configuration of sustainability considerations can bolster a firm's competitive position. The Starpharma business strategy recognizes the need to partner with the right individuals and groups that can offer access to the necessary complementary assets required for successful commercialization of the intellectual capital which Starpharma develops. Intellectual property is a key factor of competitive advantage for biotechnology companies.

An important implication for managers that has emerged from this case study is that protection of intellectual property alone through legal regimes like patents cannot lead to successful commercialization.

Intellectual property needs to be considered together with business strategy and business risk. Therefore, to become an innovative organization requires a strategy that incorporates the notion of innovation at its heart. This involves redefining customers and markets and developing technology strategies to exploit internal capabilities.

The Eurocopter case reveals that both mainstream and newstream activities need to be managed in an integrated fashion in order to sustain innovative output. Equal importance needs to be placed on both mainstream and newstream capabilities in order to initiate and sustain innovative output. Additionally, as these two streams provide complementary but interdependent resources, they need to be integrated. This means that newstream capabilities without a supporting mainstream structure are unlikely to stimulate innovation. Accordingly, innovation cannot be confined to a specific functional or positional part of the company. Rather, it has to be incorporated into an organization-wide mindset that underlies all business processes.

This is further reinforced by the Sartorius case, which demonstrates that business strategy is a major determinant for leveraging the company's innovation capabilities. Sartorius pursues a business strategy of value innovation and focuses on the anticipation of existing and potentially new customers' future needs based on the technology portfolio the company possesses or would be able to provide. At the same time, Sartorius' strategic transformation from a traditional product-focused company to a total solution provider exerts a major impact on the way the company is managed and structured.

Three major innovation capabilities were derived from the case analysis. First, Sartorius actively promotes constant intra-company knowledge diffusion as well as collaborative ties and interchange with external partners. A second, related innovation capability is the recombination of existing expertise and technologies to develop innovative solutions. For example, Sartorius has systematically exploited the experience it gained in developing required environment-friendly solutions to facilitate the development of a new business area. A third innovation capability lies in the company's effort to systematically integrate its customers into the product development process. This close interaction ensures that the innovative output is linked to a measurable value that derives from providing a fit with customer needs, thus leading to commercial success.

Although UGM has been operating without a formal innovation management model, it has managed to sustain a strategic platform that has enabled synergy between all the elements of the innovation capabilities mix such as leadership, strategy and people competence. Similarly, the DPEC case shows that innovation strategies need to be supported by knowledge management (KM) strategies so as to maximize the use of company experiences, both collectively and individually.

These strategies are necessary for the development of horizontal exchange (expanding on the existing exchanges that occur between managers) with many more levels (people, groups) so that knowledge which is embedded in products would not be lost. Invincible Company developed innovation capability through strategies that involved understanding the emergence of innovative ideas and the evolution of specific practices in managing innovation capability. Key success factors to innovation included approaches such as the "we do better" strategy, which replaced traditional processes with more supportive ones.

Furthermore, the strategic choice made by Sun Microsystems (SMS) in late 2002 to add a line of cheaper servers based on Intel chips had a significant impact on the firm's ability to maintain its current R&D spending levels, which in turn had implications regarding its ability to compete on the basis of higher-performing solutions. The Vaisala case supports the conceptual argument in the literature that all three factors of NPD, SDO and e-Commerce exert an influence on a firm's innovation capabilities and that their integration as part of the organization's business strategy is likely to increase innovation performance. Furthermore, additional capabilities such as an aligned organizational structure, a highly skilled workforce and a high research and knowledge intensity play important roles in supporting the use of these drivers.

Synthesis

Development of innovation capability requires a strategy that incorporates the notion of innovation at its heart. This involves redefining customers and markets and developing technology strategies to exploit internal capabilities. Both mainstream and newstream activities need to be managed in an integrated fashion in order to sustain innovative output. Equal importance needs to be placed on both mainstream and newstream capabilities in order to initiate and sustain innovative output. Additionally, as these two streams provide complementary but interdependent resources, they need to be integrated. Innovation-driven organizations see business strategy as a major determinant for leveraging the company's innovation capabilities. Innovative organizations go through a strategic transformation from a traditional product-focused company to a total solution provider, which in turn exerts a major impact on the way the company is managed and structured.

Innovative organizations sustain a strategic platform that enables synergy between all the elements of the innovation capabilities mix such as leadership, strategy and people competence. Innovative organizations support the conceptual argument that all three factors of NPD, SDO and e-Commerce seem to exert an influence on a firm's innovation capabilities, and that their integration as part of the organization's business strategy increases innovation performance. Furthermore, additional capabilities, such as an aligned organizational structure, a highly skilled workforce and a high research and knowledge intensity play important roles in supporting the use of these drivers.

11.3 Harnessing the Competence Base

The competence base can be harnessed by a supporting human resource infrastructure that facilitates employee empowerment, knowledge-sharing and teamwork. These are critical drivers for an innovative organization. For internal work practices to be effective, the human resource strategy of the organization must include the need to develop employees as innovative people. In essence, employee management must fit with the organization's strategy. This is reflected in the Sartorius case study. Sartorius developed innovative solutions that anticipate potential customer needs, which require a highly qualified and innovative workforce. Wright and Snell (1998) advise that organizations need to align human resource practices, employee skills and employee behaviors.

Thus, an effective training program can act to embed the necessary skills and behaviors required by the business strategy. The success of innovating the core business processes and establishing systematic customer integration at Sartorius was, to a considerable extent, dependent upon contingency factors such as work environment and employee involvement. The Sartorius College was a major facilitator in this regard. In addition, the company retains its experienced and innovative employees by providing intrinsic and extrinsic rewards. In particular, the flexible job design prevalent at Sartorius enables its employees to accompany innovation projects from the development of the innovative ideas until their commercialization. This is an important motivator to retain the company's innovation champions in the long-run, thereby ensuring a reliable staff base for initiating new innovations (Cappelli, 2000).

An important implication of nurturing this knowledge culture is the notion of knowledge exchange. In this respect, scholars highlight that employees do not only increase their knowledge base through company input in the form of training and development, but also through the systematic sharing of individual knowledge among colleagues, which in turn can be an important determinant of competitive advantage (Argote & Ingram, 2000). Also, research suggests that access to heterogeneous knowledge within the company is crucial for innovativeness (Rodan & Galunic, 2004). In this regard, Eurocopter facilitates communication, interaction and support between the various functional areas.

Additionally, by leveraging individual knowledge and sharing it among the workforce, the company was able to contain the loss of unique knowledge that occurs through employee turnover. Eurocopter also facilitated the change of mindset among employees by creating team spirit and systematic job rotation. This has increased both the flexibility of allocating human resources as well as employees' extended understanding of overarching corporate processes. The latter is particularly instrumental with regards to fostering creativity and generating ideas with the ultimate aim of initiating innovative output (Lawson & Samson, 2001).

The competence base of the people working at UGM has a direct relationship to corporate performance and innovation. This is why the organization has invested in establishing and running Caterpillar University in the USA and a training center in Australia. UGM rotates people on different projects, thus enabling them to multi-skill and to have a better appreciation of how the system works with all its interconnected and interdependent parts. In this context, innovation performance and the competence base of the workforce appear to have a strong and positive relationship.

Creativity and ideas management do not just happen by chance. Rather, they are a product of a system that encourages innovation and entrepreneurship and provides resources for people to engage in critical thinking. UGM's focus on creating and sustaining a global niche dominance in underground mining machinery has been supported by a system and an environment that nurtures creativity and ideas management. The integration of continuous customer feedback and market intelligence in the creation of innovative new products has been instrumental in facilitating continuous innovation at UGM. The promotion of a knowledge-sharing culture, based on rotation of staff on diverse new product development projects and the use of information technology, has enabled UGM to profit from creativity and ideas management.

In terms of the breadth and depth of skill within the field of Client Solutions, SMS has a small group for each product that does all the thinking in terms of product planning. SMS also has a so called "engine room" that looks at processes and how to match that up with the products:

We are looking at how to push out to market solutions, not products.

Sustainable development considerations are also reflected in the organization's culture by way of shared values and fostering an environment where the people within the organization are encouraged to be innovative. The organization offers formal and informal rewards for innovative contributions. Any employee in the organization whose ideas are adopted and shows a return to SMS receives financial rewards and is recognized as part of the formal performance review process.

In a similar fashion, DPEC harnesses the competence base by using small teams which cover several business areas. While not specifically charged with being innovative (they are actually charged with providing solutions within their R&D budget and timelines), innovative strategies did emerge out of solution-focused approaches, because by working as focused teams, they knew their own work and each others'.

Vaisala's operations are based on core competencies that the company continues to develop. It maintains a high knowledge and research intensity that translates into superior innovative outputs which have made the company a leader in technological progress in its immediate field of operational focus. Vaisala has a high skill and competence base, with 41 percent of its global workforce holding a university degree. The resulting heterogeneity of knowledge residing within the firm's employees (Grant, 1996) is thought to lead to superior innovation performance (Rodan & Galunic, 2004). Vaisala further distinguishes itself from its main competitors by offering a comprehensive range of applications and services with regards to environmental measurement and meteorology, thereby serving as a total solution provider for its customers. This approach provides the company with a substantial differentiation advantage.

The workforce at Starpharma is also highly educated, with 11 employees holding PhD qualifications. The senior management team, representing one-third of the organization, can be characterized as possessing a multi-disciplinary outlook and strong appreciation of technology, with almost all members holding at least one qualification in the field of science. The small size of the workforce is conducive to fostering heightened awareness in all employees of the business and activities in which Starpharma is involved. Roles and responsibilities for individuals within the organization are broad, and not limited to specific technical functions — this is a product of the culture at Starpharma. An example of the culture can be observed by the response of the Intellectual Property and Commercialization Manager upon discussing formal procedures for managing intellectual property.

Invincible Company harnesses its competence base by establishing a formal system of annual evaluation of staff performances for its 21 employees. This was a relatively simple program agreed to by the management board, by which employees who satisfied work targets received salary/wage increases. Bonuses were also awarded to all employees at the end of each year as a result of the company's profits. However, this system was adjusted to recognize and reward innovative suggestions made by staff about the development process of the new firefighting trucks. The recognition and reward were not fixed or formal, but determined by the manager, depending on the impacts of the ideas on the project.

Synthesis

Multiple cross-case analysis (MCCA) revealed that for organizations to harness their competence base, its employees need to be close to the market in order to understand customer processes and their requirements. This entails the need to continuously train and develop the workforce, readjust job profiles and enable employees to rotate between different jobs and processes to facilitate multi-skilling. Several of the case study companies addressed this requirement by actively promoting a knowledge-sharing culture, based on the belief that innovative activity involves the need to facilitate knowledge exchange between organizational members. Various HR tools such as empowerment were used by the case study companies to initiate far-reaching change in the mindset of their employees from a functional focus towards integration and customer orientation. Generally, case study companies harnessed their competence base by establishing formal recognition and reward systems linked to the annual evaluation of staff performances. These systems were adjusted to recognize and reward innovative suggestions made by staff.

11.4 Leveraging Information and Organizational Intelligence — Absorptive Capacity

Information and organizational intelligence can be leveraged by developing absorptive capacity, which is defined by Cohen and Levinthal (1990, p. 128) as:

> ... the ability of a firm to recognize the value of new, external information, assimilate it, and apply it to commercial ends.

A strong body of research highlights the role of formal collaborative ties with external partners in terms of increasing the innovation output of biotechnology firms. Existing research (e.g., Longenecker & Meade, 1995) confirms that local knowledge of the respective customer contexts and close customer relationships are main success factors for sustained customer service.

In the case of Sartorius, the collaboration with external partners is of central importance for the development of innovations. In particular, the firm, in line with other biotechnology companies, maintains a close relationship with scientific knowledge sources. The scope of external collaborations at Sartorius encompasses a wide array of activities. For example, the company focuses on close cooperation with centers of excellence in the respective scientific field, and actively contributes to the creation of new centers of excellence, finances PhD theses, organizes seminars and workshops, hosts conferences and congresses, maintains a dialogue with scientists on a regular basis and assigns contract work to certain universities.

At Eurocopter, the use of strategic networks and partners occurs along three main dimensions. First, the company is involved in various joint manufacturing projects with other companies. The collaboration is considered to have resulted in the helicopter model featuring superior technology, reliability and flight performance. Second, Eurocopter maintains joint R&D facilities and collaboration with other industrial partners in the aerospace industry. This approach emphasizes the need to implement an appropriate formal business structure that is conducive to simultaneous innovation processes (Burgelman & Maidique, 1988). Third, Eurocopter places great importance on close collaboration with its suppliers in the design process of blades. For example, joint discussions with suppliers help to clarify which composite materials are to be used and which part of the development process can be outsourced to suppliers. This ensures that the final product provides leading-edge quality to the customer. In doing so, the company aims to create a win-win situation where close collaboration benefits both Eurocopter and its suppliers by achieving synergies, continuous feedback and long-term cooperation.

At UGM, product development managers are responsible for external collaboration. UGM collaborates with suppliers and customers to discuss the performance of existing and the characteristics of new products. UGM has also been involved with a Brisbane-based Cooperative Research Center of Excellence in Mining. The scope of UGM's collaborative effort extends to working with the CSIRO's former Mining Division, the Australian Metal Industries Research Association and a small technology company that is working on automation products.

One of the hardest challenges facing SMS is maintaining its ability to continuously absorb and act on information and knowledge from the external environment. For example, the strategic information generated by Six Sigma Black Belts is used to drive the sales process. In order to maintain the innovation drive, SMS invests in assets and maintains good facilities in terms of having access to laboratories and work centers. SMS drives and implements projects with the goal of creating a win-win situation for the customers. Occasionally, some of the other partners have an idea which they are running and SMS helps them with its facilities. Sometimes, a customer may require SMS's assistance with a go-to-market strategy. SMS has six Global Art Wall Centers, which are innovation environments where customers, partners and SMS can put together a solutions testamur and use it for marketing. SMS uses a two-plus-two approach to absorb information on innovation from the external environment. Generally, it goes through a qualification process which consists of a two-hour presentation to see if an idea gets some traction. If the idea gets some traction, then SMS will do a two-day workshop and flash out a proof of concept. From that point, SMS will do a two-month or two-year proof of concept, whatever it turns out to be. It is taken in stages and is a spaced approach so SMS can both manage risk and get the buy end.

External influences for DPEC were mainly the use of the organizational internet-based technology in its innovation with customers, industry partners and suppliers. For example, the internet and intranet were used by teams to do research, product comparisons for market analysis, and generally communicate ideas, concepts and information. The speed at which information was exchanged contributed to defining a niche market space. For example, the company was able to commercialize technology which evolved from exchanges between teams of external research and technology organizations.

Vaisala maintains a strong customer orientation strategy in order to customize its innovative applications to its customers' present and future needs. In addition, Vaisala strives to provide comprehensive customer support, which is considered an essential part of the Vaisala service concept and encompasses maintenance, training and calibration services on a global scale. Vaisala leverages its corporate skill base through a wide array of external collaborations with specialists, research institutions and joint venture partners. By complementing and extending the internal skill base and expertise, Vaisala is able to increase its absorptive capacity and the resulting knowledge intensity for the development of superior innovations (Cohen & Levinthal, 1990; Zahra & George, 2002).

Starpharma has an enviable absorptive capacity record in relation to leveraging of technical and market expertise. The company is actively involved in acquiring knowledge and makes contributions to the growing field of dendrimer science. This is a considerably altruistic view, one that advocates cooperative effort rather than doing it alone. Executive and technical members alike are active participants, formally and informally, within scientific collaboration partnerships, alliances and the wider pharmaceutical community. Such networking opportunities are of importance to assist Starpharma in understanding emergent trends in the market and to be able to evaluate and direct its resources to the projects that offer best success opportunities. Starpharma has a broad range of channels through which the company sources intellectual property. Starpharma's global research network, spanning widely across Australian, American and European research groups, affords the company similar influence to generate relevant intellectual property.

Intarakumnerd *et al.* (2002) considered that a national innovation system in Thailand results from strong links between firms, universities and government organizations. However, individual companies (in particular, SMEs) have very different motivations and experiences which influence their pathways to innovation capability. The use of ICT, primarily the internet, has improved business competitiveness and opportunities for SMEs in Thailand, allowing them to compete on equal terms with larger organizations. The internet helps reduce costs of accurate global information exchange, contributing to the innovation capability being carried out within individual companies in Thailand (Haynes *et al.*, 1998).

Synthesis

A strong body of research highlights the crucial role of formal collaborative ties with external partners in terms of increasing the innovation output. MCCA revealed that collaboration with external partners is of central importance for the creation of innovation-driven organizations. However, one of the most difficult challenges facing innovative organizations is maintaining their ability to continuously absorb and act on information and knowledge from the external environment. The scope of external collaborations in innovation-driven organizations encompasses a wide array of activities such as close cooperation with centers of excellence, financing of PhD scholarships, organizing seminars and workshops, hosting conferences and congresses, maintaining dialogues with scientists on a regular basis and assigning contracts for collaborative work with certain universities.

This approach emphasizes the need to implement an appropriate formal business structure that is conducive to simultaneous innovation processes. Innovation-driven organizations place great importance on close collaboration with their suppliers. This ensures that the final product provides leading-edge quality to the customer. In doing so, innovation-driven companies aim to create a win-win situation in which close collaboration benefits both the organization and its suppliers by achieving synergies, continuous feedback and long-term cooperation.

Product development managers in innovation-driven organizations are typically responsible for external collaboration with suppliers and customers to discuss the performance of existing and the characteristics of new products. Innovation-driven organizations use the internet and intranet for teams to do research, product comparisons for market analysis, and generally communicate ideas, concepts and information. This allows organizations to maintain a strong customer orientation in order to customize their innovative applications to customers' present and future needs.

Networking opportunities are of importance to aid innovation-driven organizations in understanding emergent trends in the market and to evaluate and direct resources towards best success opportunities. The use of ICT, primarily the internet, has improved business competitiveness and collaborative opportunities for innovation-driven organizations by reducing costs for accurate global information exchange, thus contributing to the innovation capability being carried out within innovation-driven organizations.

11.5 Possessing a Market and Customer Orientation

Making sure customers' needs act as the prime driver for innovation is deemed to be a critical issue in all case study companies. For example, Sartorius has established a mission that explicitly states its focus on innovation and technology orientation. Likewise, the company has systematically shifted its overall competence from product orientation to total solution management, enabling customers to implement complex processes both in a laboratory and a production environment. In doing so, the company aligns its strategic activities towards constantly creating customer value.

Embarking on a large-scale process innovation project, Eurocopter was able to reengineer and realign its core business processes. The example of the blade repair process shows how the company moved from a sequential vertical structure to a horizontal approach with continuous customer interaction. This customer involvement, in turn, became part of a more far-reaching change in customer philosophy at Eurocopter. The case demonstrates how a systematic integration of customers into the product development process provided the company with a key capability to initiate innovations by creating new and sustained customer value.

Delivering customer value is a key driver throughout UGM's operation. Continuing research into customer needs and the value it places on products and services help the organization form a clear vision of industry direction. Bringing together a focus on customer needs and the increasing use of sophisticated technology, engineering, continuous improvement and R&D provide the synergy to develop and implement appropriate solutions.

SMS is very much a virtual organization that is influenced by both external and internal forces. The key external factors that influence the organization are government regulations, competitors and customers. SMS's sophisticated and demanding customers are focused on value and how to increase their own competitiveness. The key internal factors that influence SMS's sustainable development practices have been its CEO and senior management, who have a very strong vision for the organization. These internal factors drive the R&D program and make decisions concerning the firm's environmental performance.

Vaisala maintains a strong customer orientation in order to customize its innovative applications to its customers' present and future needs. In addition, Vaisala strives to provide comprehensive customer support, which is considered an essential part of the Vaisala service concept and encompasses maintenance, training and calibration services on a global scale.

Invincible Company possessed a market and customer orientation by having a flexible organizational structure. This allowed work practices to be a successful stimulus for the development of innovation capability. Interestingly, the manager reverted, on occasions, to more formal strategies of regular reviews of production processes, checks on compliance with customers' requirements and tight work schedules, but retaining the flexible work structure. This freedom allowed staff to consider customers' preferences, assess changes in the firefighting truck industry and make crucial decisions.

The marketing of products and services is an activity of significant priority for Starpharma's executive team. In addition to the promotion of new product development initiatives, marketing efforts are focused on promoting Starpharma's technological and commercialization capabilities to attract the interest of prospective complementary partners like "Big Pharma" with significant mass-market marketing and distribution assets. Furthermore, marketing is seen to be a function performed implicitly by all members representing Starpharma publicly in events such as international conferences.

Synthesis

Sophisticated and demanding customers are focused on value and how to increase their own competitiveness. Existing research (e.g., Longenecker & Meade, 1995) confirms that local knowledge of the respective customer contexts and close customer relationships are main success factors for sustained customer service. Making sure customers' needs act as the prime driver for innovation is deemed to be a critical issue in all case study companies. Systematic integration of customers into the product development process provides the company with a key capability to initiate innovations by creating new and sustained customer value. Delivering customer value is a key driver of innovation-driven companies. Continuing research into customer needs and the value placed on products and services help the organization form a clear vision of industry direction. Bringing together a focus on customer needs, and the increasing use of sophisticated technology, engineering, continuous improvement, and R&D, provides the synergy to develop and implement appropriate solutions. Therefore, maintaining a strong customer orientation in order to customize innovative applications to customers' present and future needs is a key driver of innovative organizations. A flexible organizational structure allows work practices to be a successful stimulus for the development of innovation capability. Flexibility of work structures and practices allow "best practice" innovative organizations to consider customers' preferences, assess changes and make crucial decisions in delivering customer value.

11.6 Creativity and Idea and Knowledge Management

Having people with the technical and professional knowledge, keeping knowledge in-house and being able to leverage from it by sharing it are important drivers of innovative organizations. Damanpour (1991) found that the technical and professional knowledge of the people in an organization are positively linked to innovation. "Technical knowledge" reflects the organization's technical resources and technical potential, while "professional knowledge" reflects both education and experience (Damanpour, 1991).

A central innovation capability can be identified in Sartorius' efforts to foster internal diffusion and external exchange of knowledge. To enhance this transfer, the company opened a corporate university called Sartorius College in the year 2001. The concept for this college goes back to the initiative of Sartorius' former CEO, who emphasized the need for the constant sharing of knowledge in order to reap knowledge benefits in terms of innovation and continuous improvement. The construction of the Sartorius College can be viewed as a systematic investment in the company's knowledge base.

Eurocopter confirmed the need to exchange knowledge and resources across organizational boundaries. Networks and alliances with key customers, suppliers, competitors and other participants help integrate complementary innovation capabilities, thereby fostering the development of new business streams. This is particularly relevant in high-technology environments such as the aerospace industry, where firms will not be able to maintain competences in all potentially important technical areas.

At UGM, knowledge on NPD is shared with other divisions of the global Caterpillar organization through the intranet and other networking opportunities. Caterpillar's Melbourne Training Center is recognized as the best in the industry because it provides professional, leading-edge, practical training. It also demonstrates UGM's commitment to the continual development of its employees, who service the dealer network, and to the training of dealer staff, who support the widening product and customer base. The Training Center's courses provide dealership personnel with the skills and knowledge to meet the full range of customer needs.

Knowledge is stored in SMS's extensive work sites and systems. Knowledge-based centers have actually been built to service its needs on a global basis so that SMS can leverage the benefits of being a global organization.

The process of developing and applying innovation knowledge at DPEC occurred in two different phases. The first phase built up new knowledge through various innovation practices, such as the design of original products by small divisional teams. These practices reflect the customers' and other stakeholders' requirements. The second phase captured existing innovation knowledge through reflection, learning and understanding of key steps throughout the innovation process. Knowledge transfer was seen as critical in reducing the NPD lead time, so as to meet the needs of the customer and the organization in terms of time, cost and quality.

Vaisala adopts a highly decentralized approach to knowledge diffusion. Indeed, the process of lodging and accessing information over the intranet is dyadic and interactive. Through this approach, the company nurtures a knowledge-sharing culture that encourages its employees to actively diffuse individual knowledge and thus create and extend an organization's value.

The dispersed nature of Starpharma's research network relies on cost-effective internet and e-Mail-based methods to interact and communicate its discussions, designs, results and data. Internally, Starpharma's network provides multi-user access to knowledge directories, and manuals for procedures and processes. Investigation for information technology solutions for knowledge management and collaborative project management has been considered to enhance current practices.

Invincible Company's NPD process involved innovative knowledge-sharing. The transition to phase 2 began when several conceptual designs had been completed and agreed to by the customer. Three teams — design, production and procurement — were established to discuss the approach to each design and the various possibilities to produce a firefighting truck at Invincible. The meetings of these teams were frequently arranged in formal or semi-formal environments every morning for approximately two months to facilitate the knowledge-sharing process leading up to the finalization of the design or starting date for the prototype process.

Synthesis

Having people with the technical and professional knowledge, keeping knowledge in-house and being able to leverage from it by sharing it are important drivers of innovation capability. Innovative organizations establish an internal college for the constant sharing of knowledge and the systematic investment in the company's knowledge base. Networks and alliances with key customers, suppliers, competitors and other participants are able to integrate complementary innovation capabilities, thereby fostering the development of new business streams. This is particularly relevant in high-technology environments such as the aerospace industry, where firms will not be able to maintain competences in all potentially important technical areas. Innovative organizations establish knowledge-based centers to service their needs on a global basis so that they can leverage the benefits of being a global organization.

The process of developing and applying innovation knowledge can occur in two different phases. The first phase builds up new knowledge through various innovation practices, such as the design of original products by small divisional teams. These practices reflect the customers' and other stakeholders' requirements. The second phase captures existing innovation knowledge through reflection, learning and understanding of key steps throughout the process of innovation.

Knowledge transfer is seen as critical in reducing the NPD lead time so as to meet the needs of the customer and the organization in terms of time, cost and quality. Innovation-driven organizations nurture a knowledge-sharing culture that encourages their employees to actively diffuse individual knowledge. The dispersed nature of

Starpharma's research network relies on cost-effective internet and e-Mail-based methods to interact and communicate, for discussions, designs, results and data. Innovation-driven organizations have internal networks which provide multi-user access to knowledge directories, and manuals for procedures and processes.

11.7 Organizational Structures and Systems

Organizational structures and systems can help or hinder the innovation performance of the organization. Innovative organizations ensure that their internal organization translates such inputs into specific organizational actions and plans (Ahmed, 1998). Internal organization includes an organization's structure, the processes it employs and its people. Underpinning these aspects is the culture that drives innovation in every aspect. Innovative organizations continually improve their business performance and on-going viability by integrating the latest approaches to managing the flow of work.

This includes the adoption of advanced technologies, new work methods and redesigning commercial relations with information technology. Such organizations focus on quality and cycle time and reap the rewards with increased market share and financial performance. Eurocopter embarked on a large-scale process innovation project and reengineered and realigned its core business processes. The example of the blade repair process shows how the company moved from a sequential vertical structure to a horizontal approach with continuous customer interaction. This resulted in decreased cycle time in the blade repair process.

Sartorius developed a new group-wide strategy focusing on growth and innovation. Sartorius strategically aimed at expanding its technology and product portfolio by using its sales subsidiaries and local commercial agencies more efficiently and developing the business through organic sales increases and acquisitions. In the following years, the firm acquired various companies to strengthen the strategic portfolio and reorganize the company to its present structure. At the same time, the innovation rate picked up and the company managed to increase the scope of its innovation in terms of commercial success.

UGM has a flat organizational structure and its systems for supporting creativity and ideas management have played a positive role on the innovation performance of the organization as measured by various innovation metrics. Innovation is measured by commercial success, which depends on how customer-focused the organization is and whether it is coming up with innovations that customers want.

The internet has forced some organizations to review existing processes and practices and to reconfigure their innovation capabilities. SMS is a good example of how internet-based technology has influenced the innovation capability of the firm. Internet-based technology is part of SMS's strategy in the way employees work and how the organization is virtually structured. Furthermore, the intranet has provided flexibility, security and the ability to retain knowledge.

Innovation can also occur on a large scale in an organization such as DPEC, which is comprised of several main divisions. DPEC is primarily hierarchical and has autonomous business units within the divisions. Interaction occurs through various functional areas. For example, project management skills, finance skills and R&D management are carried out across all the divisions.

Electronic information exchange at Vaisala enables joint product development through the use of virtual teams across organizational and geographical units, pointing to another benefit of internet-based technology (Salazar *et al.*, 2003). Vaisala adopts a highly decentralized approach to knowledge diffusion. Indeed, the process of lodging and accessing information over the intranet is dyadic and interactive. Through this type of structure, the company nurtures knowledge-sharing that encourages its employees to actively diffuse individual knowledge and thus create and extend an organizationally valuable knowledge base. Vaisala emphasized that an internet-based structure increases efficiency and helps to allocate resources to the innovation process.

Formal systems and information technology are enablers in the daily operations at Starpharma, whose activities are structured around pharmaceutical product development. Therefore, strict practices and compliance with regulatory agencies are required. Starpharma's implementation of a quality management system has been specifically developed in compliance with international standards and regulations. Implemented quality systems also make use of information technologies, of which the assurance of data integrity is a high priority. Starpharma believes that data accuracy and integrity is critical to reducing product development time. Internally, Starpharma's network provides multi-user access to knowledge directories, and manuals for procedures and processes. Investigation for information technology solutions for knowledge management and collaborative project management has been considered to enhance current practices.

At Invincible, as part of its "we do better" strategy, management formulated many new activities and procedures that tended to be flexible and informal. These were apparent in the organizational structure, especially for the staff in the engineering design team where several new joint-teams, including staff from different units, were established. For the design team, engineers with a range of expertise, qualifications and specializations formed one large team. Team members were not representative of any particular unit, but acted in co-coordinator roles to provide diverse input for the new product designs. Generally, the team discussed and commented on improvements to outperform its competitors.

Synthesis

Innovative organizations move from a sequential vertical structure to a horizontal approach with continuous customer interaction and focus on growth and innovation, and have systems for supporting creativity and ideas management. Internet-based technology has influenced the innovation capability of innovative organizations. Furthermore, the intranet has provided flexibility, security and the ability to retain knowledge. Innovation exists as part of a standard procedure, which eventually leads to the development of a product for customers. Innovative organizations adopt a highly decentralized approach to knowledge diffusion. Electronic information exchange enables joint product development through the use of virtual teams across organizational and geographical units, which points to another benefit of internet-based technology.

Through this type of structure, the company nurtures knowledge-sharing that encourages its employees to actively diffuse individual knowledge and thus create and extend an organizationally valuable knowledge base. Innovative organizations have internal networks and systems which provide multi-user access to knowledge directories, and manuals for procedures and processes. Implementation of quality management systems and information technologies provide data accuracy and integrity, which are critical to reducing product development time. Innovative organizations have a dual-type organizational structure: flexible but sometimes formal, driven by the requirements of different types of projects, products, development phases and/or production stages.

11.8 Culture and Climate

Development of innovation capability in organizations is strongly reliant on a strong innovation culture. Effective human resource strategies ensure that employees are acting in concert with the requirements of the organization. However, to sustain such behavior and propel the organization to world class, the work environment or work culture needs to be motivational (Miles, 2000).

Sartorius has addressed this requirement by actively promoting a knowledge-sharing culture, based on the belief that innovative activity involves the need to facilitate knowledge exchange between organizational members. Since knowledge primarily resides within the individual (Grant, 1996), employees have to be motivated to share individual knowledge. Sartorius enables its employees to accompany innovation projects from the development of the innovative ideas until their commercialization. This is an important motivator to retain the company's innovation champions in the long-run, thereby ensuring a reliable staff base for initiating new innovations (Cappelli, 2000).

At Eurocopter, an innovation culture has evolved in the form of a state-of-mind that guides both employees and corporate decision-making. Innovation encompasses both process and product innovation. From a process perspective, innovation occurs in terms of a reorganization of work processes. From a product perspective, innovation is perceived as the outcome of application research that ultimately leads to innovative products or parts development. Eurocopter makes extensive use of cross-functional teams in order to stimulate both process and product innovation. In this regard, staff from the customer support side is linked up with employees from the product design and quality departments.

At UGM, the culture and climate have had a decisive impact on the organization's innovation performance. Its culture and climate have been influenced by the entrepreneurial and innovative approach of the parent company that creates conditions for creative thinking, risk-taking and a strong focus on markets and customers. The culture and climate for innovation and entrepreneurship have also been influenced by the Australian CEO and senior management, who recognize the importance and linkages between culture, climate and innovation performance. The use of Six Sigma and the balanced scorecard methods have also contributed to a culture of accountability, transparency and recognition of performance.

SMS acknowledged that its culture is underpinned by the values and mindset of its people. Getting commitment from employees is the biggest challenge. There appears to be a significant and positive relationship between the quality of the people and quality of outcomes. Sustainable development considerations are also reflected in the organization's culture by way of shared values and fostering an environment where the people within the organization are encouraged to be innovative.

Similarly, at DPEC, innovation is part of the corporate governance and the company's core values and principles. DPEC primarily utilizes small divisional teams to conduct its work within a project framework. These teams are often made up of the same people with overlapping responsibilities. These teams could be seen as the building blocks of DPEC's culture and are significant with regards to developing innovation capability.

Vaisala's culture is driven by six core values, namely science-based innovation, fair play, "one for all — all for one", customer focus, personal growth and goal orientation. These values reflect a corporate philosophy that builds upon a team-based, people-orientated and customer-focused approach to innovation.

Starpharma has a broad view of intellectual property encompassing codified and tacit knowledge as well as people and relationships, which all simultaneously contribute to the development of innovation capability. Therefore, the company believes that it is important for its corporate culture to be diffused with values based on intellectual property. Invincible Company also developed an innovation culture through strategies that involved understanding the emergence of innovative ideas and the evolution of specific practices in managing innovation capability.

Synthesis

Innovative organizations promote a knowledge-sharing culture based on the belief that innovative activity involves the need to facilitate knowledge exchange between organizational members. Innovation becomes a state-of-mind that guides both employees and corporate decision-making and encompasses both process and product innovation. Innovative organizations make extensive use of cross-functional teams in order to stimulate both process and product innovation. The culture and climate of innovative organizations are influenced by the entrepreneurial and innovative approach of the parent company that creates conditions for creative thinking, risk-taking and a strong focus on markets and customers.

The culture and climate for innovation and entrepreneurship have also been influenced by the CEO and senior management, who recognize the importance and linkages between culture and climate and innovation performance. In sum, the culture in innovation-driven organizations needs to encourage innovation by explicitly recognizing employees' efforts and being prepared to take risks. Innovative organizations actively introduce change in the form of innovative work structures, work processes and employee management, and generate a culture that allows the entrepreneurial spirit to thrive in every facet.

11.9 Management of Technology and Its Use

It is often stated that we live in the knowledge era. The web has transformed our lives in many ways, one of which is to flood us with available information. Distilling such information is vital if an organization wishes to interpret its meaning and relevance. Innovative organizations understand this need and have established competitive intelligence. For example, Sartorius has established a mission that explicitly states its focus on innovation and technology orientation. The company has systematically shifted its overall competence from product orientation to total solution management, enabling customers to implement complex processes both in a laboratory and a production environment. In doing so, the company aligns its strategic activities towards constantly creating customer value. The company's mission statement reflects a corporate philosophy that views innovation as an integral part of conducting business. As a result, the company does not maintain a separate department for innovation, but rather integrates innovation as an overarching concept into the everyday processes.

A further example is provided by Eurocopter, which has developed its own technology with regards to manufacturing and repairing rotor blades and related products. The design and manufacturing stage are closely linked as an integrated process that incorporates computer-aided design, large-scale production of machining and core materials, the application of leading-edge technology to manufacture complex component parts and extensive test facilities. In a similar approach to Sartorius, Eurocopter does not locate innovation in a specific department, but rather sees it as a process that is prevalent in everyday routines. As a result, innovation becomes a state-of-mind that guides both employees and corporate decision-making.

UGM has strong values that are based on providing innovative products and services and a passion for satisfying customers. UGM's core business relies on the management of technology and its application to meet customer needs. UGM's ability to customize its products and services to suit customer needs is highly dependent on its innovation culture, which is also driven by technology and its application. Continuous research, training and education have enabled employees to develop and maintain the necessary skills and motivation to manage the innovation process and to make the most of existing and emerging technologies.

SMS is also a technology-centered organization. Its professional services experts have provided single-point-of-contact solutions to fit business needs. Given that SMS is an internet-based organization, around 80 percent of SMS's business comes from large corporate organizations and government who are mature e-Commerce users. Its hardware is increasing in sales exponentially. However, according to the state manager,

> If we stay in the hardware business only, we are going to go out of business.

e-Business enables SMS to extend its range of services and thus to position itself as solving customers' problems, not just providing "off the shelf" technology. The message is loud and clear from several of the case study companies that technology should be the ingredient for the provision of integrated solutions for customers and not to be seen as an end in itself.

DPEC is a technologically diversified organization. The key message from the DPEC case study is that technology combined with the appropriate application knowledge plays a key role in the organization's success in delivering projects on time, within the budget and following customer specifications.

Vaisala perceives technology as an enabler of value innovation. The company maintains a high knowledge and research intensity in order to be able to effectively recombine existing knowledge, create marketable research outcomes and translate the resulting new applications into the firm's product portfolio. In doing so, the company ensures that sustainability and new product development are constantly linked to leading-edge technology and are able to provide a differentiating advantage that can continuously drive innovation.

An important implication for managers that has emerged from the Starpharma case study is that protection of intellectual property alone through legal regimes like patents cannot lead to successful commercialization of technology. Investment into legal claims over intellectual property needs to be considered together with business strategy and business risk.

The Invincible case study reveals that staff with advanced technical knowledge provided accurate and detailed information for the required technical specifications of the product, which enabled the design team at Invincible to refine the conceptual product design. Product design requirements originated from the customers needing to increase the efficiency of the firefighting trucks, but the final product concepts were heavily influenced by the Invincible manager and his staff. The implication from this case study is that collaboration of customers' clear requirements and the ability of the Invincible staff to bring these requirements to reality through technology management and application is a critical step in the commercialization process.

Synthesis

Innovative organizations understand the need for managing knowledge and have established competitive intelligence or information technologist functions within their organizations. All case study companies have systematically shifted their thinking from product orientation to total solution management, enabling customers to implement complex processes towards constantly creating customer value. Case study companies do not maintain a separate department for innovation, but rather integrate innovation as an overarching concept into the everyday processes. As a result, innovation becomes a state-of-mind that guides both employees and corporate decision-making.

The ability of innovative organizations to customize their products and services to suit customer needs is highly dependent on their innovation culture, which in turn is driven by technology and its application. Furthermore, technology combined with the appropriate application knowledge plays a key role in the organization's success in delivering projects on time, within the budget and following customer specifications. Innovative organizations maintain a high knowledge and research intensity in order to effectively recombine existing knowledge, create marketable research outcomes and translate the resulting new applications into the firm's product portfolio.

In doing so, innovative organizations are linked to leading-edge technology and are able to provide innovation capability that can continuously drive innovation. This also entails the need to access knowledge from external sources to complement the corporate skill base, and to maintain a diversified workforce in order to increase the organizational absorptive capacity, thereby allowing for a higher level of knowledge to be processed, reconfigured and leveraged.

Innovative organizations focus on the protection of their intellectual property. However, protection alone through legal regimes like patents cannot lead to successful commercialization of technology. Investment into legal claims over intellectual property needs to be considered together with business strategy and business risk. In sum, innovative organizations collaborate with customers to establish clear customer requirements and then bring these requirements to reality through technology management and application. This is seen as a critical step in the commercialization process.

11.10 New Product Development (NPD)

Empirical work on NPD has focused on the relationships between various success factors, including new product strategies, and performance measures and risk (e.g., Cooper & Kleinschmidt, 1996; Firth & Narayanan, 1996). As a result, we know for example that firms that emphasized market innovativeness in their product introductions enjoyed higher returns than those that did not (Firth & Narayanan, 1996, p. 334). Sartorius' strategy of being a total solution provider entails the need to focus on optimizing customer processes. Sartorius integrates its customers into the NPD process. This ensures that innovative solutions are convertible into commercial success. The development of new products thus results from specific process-related requirements.

NPD at Sartorius is therefore very much driven by the total outcome in terms of innovative technologies that are used by customers in a combined way. Sartorius fosters NPD through a recombination of available technologies in order to offer innovative solutions. This ability has resulted from the firm's strategic focus on innovation through anticipation of its customers' potential future needs. Empirical research corroborates these findings by showing that customer orientation in innovation projects has a positive influence on NPD success and that the impact increases with the degree of product innovativeness (Salomo *et al.*, 2003).

While mainstream activities provide the necessary stability to maintain organizational functioning through process innovation at Eurocopter, the newstream activities introduce a dynamic context that requires continuous NPD as well as knowledge creation, application and recombination, and in doing so, make the organization a constantly moving target to competitors (Kiernan, 1996). Building on these ideas, Lawson and Samson (2001) argue that mainstream and newstream activities need to be managed in an integrative manner to achieve innovativeness. This can be achieved by creating innovation capabilities which, in turn, are able to combine key resources and capabilities to initiate innovation (Fuchs *et al.*, 2000).

A strategic challenge facing UGM is how to accelerate the NPD process without undermining the quality of its output. This, however, is not a pressing issue at the moment because the firm is struggling to meet a backlog of orders which is hampered by a national skills shortage, rising price and availability of steel and the desire to retain existing customers. The organization has in fact had to slow down the NPD process to allow for a return on investment from innovations. It has been using Six Sigma as a platform for NPD as well as multifunctional teams to work on a rotation basis in the NPD area.

SMS has successfully used innovation centers that focus on accelerating NPD in its organization. The innovation centers are made up of business people who are part of teams focused on continuous leveraging. SMS has a strategy for NPD and a dedicated department for innovation. The company makes use of cross-functional teams as part of the NPD process. Customers and suppliers are regularly consulted on how SMS products can better meet their needs. Branding plays an important role in facilitating the introduction of new products onto local and international markets.

At DPEC, innovation exists as part of a standard procedure, embedded in NPD, which eventually leads to the development of a product for its customer. Therefore, if the product is successfully developed and delivered, the innovation process has also been successfully developed and delivered. DPEC uses e-Communication to help product development in terms of information access and improved communication. The impact of sustainable development on new product development is in line with required regulations and compliance. Knowledge transfer is seen as critical in reducing the NPD lead time, so as to meet the needs of the customer and the organization in terms of time, cost and quality.

The NPD process at Vaisala is primarily driven by customer demand and therefore occurs in close cooperation with the customer. Accordingly, Vaisala has established a strong level of customer involvement, which is formalized through confidentiality agreements rather than commercial contracts. Empirical research confirms these findings by demonstrating that customer orientation in innovation projects has a positive effect on NPD success and that the impact increases with the degree of product innovativeness (Salomo *et al.*, 2003).

Despite the main focus on customer involvement, Vaisala attempts to integrate the whole process of NPD by involving the supplier side and making use of cross-unit development teams that, as mentioned earlier, frequently collaborate via electronic communication channels. This integrative approach to NPD requires a harmonization of sustainability considerations between Vaisala and suppliers, and emphasizes the close linkage between both factors in terms of their effect on the firm's innovation capabilities. This integration of NPD has also resulted in an acceleration of production processes.

Starpharma has an adaptable business strategy that continuously evaluates the NPD process with the evolving nature of the external market. This adaptability is evident in Starpharma's intellectual property strategy discussed earlier in this chapter.

The Invincible case study reveals a relationship between NPD and e-Business, excluding sustainable development. The sustainable development component (such as environmental awareness) was not evident at the product conceptualization phase, nor did it emerge as a prime focus of the product innovation process. Their application appeared to be a direct determinant of the NPD process. These findings reveal that the practices in developing innovation capability remain fragmented, lacking a systemic approach. Development of more systemic practices that bring in various elements of all three components (NPD, e-Business and SDO) are required to build innovation capability for organizational long-term sustainability.

11.11 Sustainability

Sustainability has clearly begun to assert itself as a driver for innovation. Polonsky (2001) argues that "going green" provides a firm with strategic advantages including lower costs, differentiation and revitalization. Research confirms that the establishment of environmental management systems is a powerful tool for multinational companies to stay ahead of environmental regulations that differ widely across countries they operate in (Sharfman *et al.*, 2004).

Sartorius' corporate strategy reflects a responsibility towards the environment. The company has been certified in accordance with the ISO 14001 standard. In many areas, Sartorius has developed exemplary solutions that protect both the environment and resources. Furthermore, the company demonstrates an environmental consciousness in terms of building its product portfolio. Sartorius College offers courses on practical environmental protection and relays information about the environmental audit based on the European Eco-management and Audit Scheme and the certification process according to the ISO 14001 standard. There seems to be a strong external influence on the application of practices related to sustainable development at Sartorius. More specifically, legal requirements and monetary aspects affect Sartorius' environmental consciousness. However, from a customer perspective, Sartorius does not yet seem to benefit from introducing environment-friendly products into the market. Both respondents noted that customers are still mainly unwilling to pay a premium for green products.

At the same time, the knowledge the company gained from its experience in developing environmentally conscious solutions has provided the basis for building up the new business area, Environmental Technology, which is now set up under the biotechnology division. It is quite evident from the case study that Sartorius has made systematic use of its experience in developing environment-friendly products and processes, both through external initiatives in terms of legal requirements as well as internal recombination of existing knowledge, to facilitate the development of a new business area. By adding environmental innovations to its total solution approach, Sartorius acknowledges the growing importance of green product development for increased operational excellence and long-term customer acceptability. This does not only help to maintain corporate competitive advantage, but also ensures sustainable development for society.

Eurocopter maintains a wide array of R&D projects that are resourced with approximately 10 percent of annual sales. A key focus of R&D is associated with noise abatement. Other R&D projects entail work on full all-weather capability of helicopters and the potential for optimizing in-flight comfort and the safety of pilots and passengers. Finally, the company faces external pressures with regards to sustainable development. Specifically, European legislation requires the company to focus on waste minimization and the use of environmentally friendly products. This results in certain firm resources being allocated towards meeting environmental requirements.

UGM has a similar approach to sustainability, guided by its Code of Worldwide Business Conduct. These guidelines meet or exceed local environmental regulations and assist UGM to develop solutions to customers' environmental challenges, advocate free trade and take the lead in the business community on important issues.

The SMS case study shows that both e-Commerce and sustainable development are contributing factors within innovation capability. The organization has yet to take advantage of e-Aim, which is a knowledge-capturing innovation that can have a significant impact on the acceleration of NPD. e-Aim can assist with the process of developing NPD by allowing people who are working on different project teams or innovation teams to share knowledge and learn from other parts of the organization, and extend this through SMS's network and supply chain. Like in other innovation-driven organizations, e-Commerce and SD practices have facilitated the acceleration of NPD at SMS.

DPEC is active in taking care of the environment as well as integrating into the community. The company has a strong philosophy of performance in all areas, externally as well as internally, as reflected in its corporate governance. Internally, the organization's corporate governance covers sustainable development. Externally, sustainable development is primarily influenced by government regulations that cover the defence sector, which are generally considered more stringent and demanding than for many other industries. The company does not have an explicit corporate strategy on sustainable development. However, many of the relevant issues are covered by its occupational health and safety rules.

Vaisala incorporates environmental considerations into its business operations as a result of both external influences and internal drivers. On one hand, the company ensures compliance with environmental government regulations concerning an environmentally friendly conduct of business. To achieve this conformity on an on-going basis, the company has established an ISO 14001-based environmental management system that addresses and constantly monitors all key environmental indicators. Other external influences to incorporate SD aspects stem from customers and suppliers in terms of ensuring the use of compatible processes and interfaces. At the same time, environmental protection and concern are an integral part of the company's mission and are deeply embedded in the corporate philosophy. To summarize, sustainable development at Vaisala forms not only part of an on-going aim to comply with external environmental legislation but, more importantly, is deeply embedded in the corporate philosophy and translates into a key competence and enabling factor for the firm's innovation capability.

Starpharma's activities are based around pharmaceutical product development, strict practices and compliance with regulatory agencies. Its implementation of a quality management system has been specifically developed in compliance with international standards and regulations, including those defined by the United States Food and Drugs Administration (FDA), as well as the Australian Therapeutic Goods Association's (TGA) Codes for Good Manufacturing Practice (GMP) and Good Clinical Practice (GCP).

11.12 e-Commerce

e-Commerce and the use of the internet at Sartorius are of specific value in terms of both e-Procurement and as a potentially new distribution channel. In addition, international work groups at Sartorius interchange data using the company's intranet. However, e-Commerce cannot be a source for innovation in itself, but rather has to be viewed as an enabling tool that can facilitate the development of innovative solutions. The internet increases the availability of information, but it is difficult for a company to differentiate itself through the use of this information. Rather, it is important to possess a firm-specific background of skills, experience and knowledge.

This background has the potential for a sustained competitive advantage. As described earlier, the Sartorius College intends to foster systematic knowledge transfer within and beyond the company and thus aims at building up this knowledge base. This finding is supported by a growing strand of literature that views e-Commerce and the internet as an enabling technology (e.g., Porter, 2001). While the internet per se will rarely result in a competitive advantage, it provides companies with better opportunities for distinctive strategic positioning than did previous generations of information technology.

UGM has embraced e-Commerce as a new way of doing business. This is evident in the reduction in manual or paper-based systems. Therefore, e-Commerce is considered as a means of cost reduction. The main uses of the internet-based technology at UGM are the intranet, funds transfer, parts order placement, company website and data transfer. UGM believes that information technology will drive machinery development in the 21st century and Caterpillar is at the cutting-edge, continually setting and then improving benchmarks.

SMS is a good example of how internet-based technology has influenced the innovation capability of the firm. Internet-based technology is part of SMS's strategy, the way employees work and how the organization is virtually structured. Furthermore, the intranet has provided flexibility, security and the ability to retain knowledge. Overall, e-Business enables SMS to extend its range of services and thus to position itself as solving customer problems. The evidence from the literature indicates that most organizations around the world have yet to develop competencies in e-Commerce and sustainable development practices with traditional business models and approaches (Dunphy *et al.*, 2003).

DPEC uses the internet and intranet for teams to do research, product comparison for market analysis, and generally communicate ideas, concepts and information. The internal motivation for adopting internet-based technology has been the cost efficiencies. At DPEC, e-Commerce has contributed to a particular organizational culture towards real-time information communication.

Vaisala's main applications of internet-based technology in its innovation process entail a company-wide intranet and e-Procurement on a global scale. While the company has established an integration of suppliers into the corporate intranet via extranet applications, the corresponding integration of customers is still in the development stage. The adoption of internet-based technology at Vaisala has been primarily driven by the requirements of knowledge storage and diffusion across different parts of the company. This is even more important as the company has many geographically dispersed business units, which makes knowledge exchange more complex. Existing literature not only supports the notion of ease of knowledge transfer across internal organizational boundaries though intranet-based communication tools (Fulk & DeSanctis, 1995), but also emphasizes benefits of internet-based technology for the development of global strategy (Yip & Dempster, 2005). Importantly, electronic information exchange at Vaisala enables joint product development through the use of virtual teams across organizational and geographical units, which points to another benefit of internet-based technology (Salazar *et al.*, 2003).

This finding leads us to conclude that there are also substantial interaction effects between the use of e-Commerce and NPD, thereby enhancing a firm's innovation capability. Formal systems and information technology are enablers in daily operations at Starpharma. The use of information technology is a prominent tool used daily within the company. R&D activities use information technologies for data collection and analysis. The bulk of data generated by Starpharma is the output of rigorous testing at each stage of the NPD process.

11.13 Characteristics of an Innovative Organization

Vision and Strategy

Innovation-driven organizations develop innovation capability based on the formulation of business strategy underpinned by innovation. This is achieved by:

- Formulation of business strategy based on the value innovation concept, which links customer value with technology innovation (Kim and Mauborgne, 1999).
- Focusing on redefining the problem through value innovation and not just coming up with a solution through technology innovation.
- Managing mainstream and newstream activities in an integrated fashion in order to sustain innovative output.
- Going through a strategic transformation from a traditional product-focused company to a total solution provider.
- Sustaining a strategic platform to gain synergy between all the elements of the innovation capabilities mix such as leadership and strategy, and people competence.
- Integration of NPD, SDO and e-Commerce as part of business strategy to influence innovation capability.
- Aligning capabilities such as organizational structure, a highly skilled workforce and a high research and knowledge intensity to support the use of NPD, SDO and e-Commerce.

Harnessing the Competence Base

Innovation-driven organizations develop innovation capability by harnessing the competence base. This is achieved by:

- Empowering employees so that they are close to the market and gain an understanding of customers' processes and their requirements.
- Developing the workforce through training, readjusting job profiles and enabling employees to rotate between different jobs and processes to facilitate multi-skilling.
- Promoting a knowledge-sharing culture, based on the belief that innovative activity involves the need to facilitate knowledge exchange between organizational members.
- Establishing formal recognition and reward systems linked to the annual evaluation of staff performances.

Leveraging Information and Organizational Intelligence — Absorptive Capacity

Innovation-driven organizations develop innovation capability by leveraging information and organizational intelligence. This is achieved by:

- Collaborating with external partners.
- Creating a win-win situation with suppliers where close collaboration benefits both the organization and its suppliers by achieving synergies, continuous feedback and long-term cooperation.
- Seeking networking opportunities to increase understanding of emergent trends in the market and to evaluate and direct resources to projects that offer best success opportunities.

Possessing a Market and Customer Orientation

Innovation-driven organizations develop innovation capability by possessing a market and customer orientation. This is achieved by:

- Systematic integration of customers into the product development process. This provides the company with a key capability to initiate innovations by creating new and sustained customer value.

- Conducting research into customers' needs and identifying the value they place on products and services. This helps the organization to form a clear vision of industry direction.
- Maintaining a strong customer orientation in order to customize innovative applications to customers' present and future needs.
- Developing a flexible organizational structure which allows work practices to be a successful stimulus for the development of innovation capability.
- Developing flexible work structures and practices to assess customers' preferences, assess changes and make crucial decisions in delivering customer value.

Creativity and Idea and Knowledge Management

Innovation-driven organizations develop innovation capability by harnessing creativity, ideas and managing knowledge. This is achieved by:

- Having people with the technical and professional knowledge, keeping knowledge in-house and being able to leverage from it by sharing it.
- Establishing an internal college for the constant sharing of knowledge and the systematic investment in the company's knowledge base.
- Integrating networks and alliances with key customers, suppliers, competitors and other participants, thereby fostering the development of new business streams.
- Establishing knowledge-based centers that have been built to service its needs on a global basis so that it can leverage the benefits of being a global organization.
- Focusing on knowledge transfer as critical in reducing the NPD lead time, to meet the needs of the customer and the organization in terms of time, cost and quality.
- Nurturing a knowledge-sharing culture that encourages employees to actively share individual knowledge.

Organizational Structures and Systems

Innovation-driven organizations develop innovation capability through organizational structures and systems. This is achieved by:

- Moving from a sequential, vertical structure to a horizontal, process-focused approach with continuous customer interaction and a focus on growth and innovation.
- Institutionalizing systems for supporting creativity and ideas management.
- Utilizing internet-based technology to enhance innovation capability.
- Adopting a highly decentralized approach to knowledge diffusion. Electronic information exchange enables joint product development through the use of virtual teams across organizational and geographical units.
- Establishing internal networks and systems which provide multi-user access to knowledge directories, and manuals for procedures and processes.
- Implementation of quality management systems and information technologies for data accuracy and integrity, which are critical to reducing product development time.

Culture and Climate

Innovation-driven organizations develop innovation capability by developing an innovation culture and climate. This is achieved by:

- Promoting a knowledge-sharing culture based on the belief that innovative activity involves the need to facilitate knowledge exchange between organizational members. The culture encourages innovation by explicitly recognizing employees' efforts and being prepared to take risks.
- Creating a state-of-mind that guides both employees and corporate decision-makers in process and product innovation.
- Making extensive use of cross-functional teams in order to stimulate both process and product innovation.

- Creating conditions for creative thinking, risk-taking and a strong focus on markets and customers.
- Demonstrating CEO and senior management commitment to the importance and linkages between culture and climate and innovation performance.
- Introducing change in the form of innovative work structures, work processes and employee management, and generating a culture that allows the entrepreneurial spirit to thrive in every facet.

Management of Technology and Its Use

Innovation-driven organizations develop innovation capability based on the management of technology and its use. This is achieved by:

- Systematically shifting management thinking from product orientation to total solution management, enabling customers to implement complex processes towards constantly creating customer value.
- Integrating innovation as an overarching concept into the everyday processes. Innovation becomes a state-of-mind that guides both employees and corporate decision-making.
- Customizing products and services to suit customer needs.
- Combining technology with the appropriate application knowledge so as to deliver projects and new products on time, within the budget and following customer specifications.
- Maintaining a high knowledge and research intensity in order to be able to effectively recombine existing knowledge, create marketable research outcomes and translate the resulting new applications into the firm's product portfolio.
- Accessing knowledge from external sources to complement the corporate skill base and maintaining a diversified workforce in order to increase organizational absorptive capacity, thereby allowing for a higher level of knowledge to be processed, reconfigured and leveraged.
- Focusing on the protection of intellectual property and recognizing that protection alone through legal regimes like patents cannot lead to successful commercialization of technology.

- Considering intellectual property together with business strategy and business risk.
- Establishing collaborative relationships with customers and articulating clear customer requirements, and then bringing these requirements to reality through technology management and application. This is seen as a critical step in the commercialization process.

11.14 Innovation-Driven Organizations: The Role of NPD, SDO and e-Commerce

Innovation-driven organizations integrate their customers into the NPD process. This ensures that innovative solutions are convertible into commercial success. Empirical research corroborates these findings by showing that customer orientation in innovation projects has a positive influence on NPD success and that the impact increases with the degree of product innovativeness. NPD at innovation-driven organizations is very much driven by the total outcome in terms of innovative technologies that are used by customers.

A strategic challenge facing innovation-driven organizations is how to accelerate the NPD process without undermining the quality of their output. These organizations make use of cross-functional teams as part of the NPD process. Customers and suppliers are regularly consulted on how products can better meet their needs. NPD is primarily driven by customer demand and therefore occurs in close cooperation with the customer. Empirical research confirms these findings by demonstrating that customer orientation in innovation projects has a positive effect on NPD success.

Sustainability has clearly begun to assert itself as a driver for developing innovation capability. Research confirms that the establishment of environmental management systems is a powerful tool for multinational companies to stay ahead of environmental regulations that differ widely across countries they operate in. There seems to be a strong external influence on the application of practices related to sustainable development at innovation-driven organizations. More specifically, legal requirements and monetary aspects affect the environmental consciousness of innovation-driven organizations.

e-Commerce acts as both a driver and an enabler of innovation within organizations. As a driver of innovation, e-Commerce has underpinned stronger, more rapid and flexible competition, forcing firms to restructure competitive boundaries and reevaluate existing practices, products and services. As an enabler of innovation, e-Commerce provides immense scope to discard old processes, diffuse local innovations globally, remove constraints to innovation and create entirely new innovative practices and models.

11.15 Conclusion

Based on the multiple cross-case analysis, the following conclusions are articulated with respect to the research question. Development of innovation capability requires a simultaneous focus:

- Adopting a strategy that incorporates the notion of innovation at its heart. This involves redefining customers and markets and developing technology strategies to exploit internal capabilities.
- Harnessing of the competence base. This entails the need to continuously train and develop the workforce, readjust job profiles and enable employees to rotate between different jobs and processes to facilitate multi-skilling.
- Leveraging information and organizational intelligence by developing absorptive capacity through collaboration with external partners.
- Having people with the technical and professional knowledge, keeping knowledge in-house and being able to leverage from it by sharing it are important drivers of innovation capability.
- Moving from a sequential vertical structure to a horizontal approach with continuous customer interaction and focus on growth and innovation, and having systems that support creativity and ideas management.
- Promoting a knowledge-sharing culture. Innovation becomes a state-of-mind that guides both employees and corporate decision-making and encompasses both process and product innovation.
- Understanding the need for managing knowledge and establishing competitive intelligence or information technologist functions within their organizations.

- Developing innovation capability based on the management of technology and its use. This is achieved by systematically shifting management thinking from product orientation to total solution management, enabling customers to implement complex processes towards constantly creating customer value.

Innovation capability can be developed and exploited through the integration of e-Commerce, sustainable development and new product development, driven by HR policy and practice that facilitates employee empowerment, knowledge-sharing, teamwork and "top-down" and "bottom-up" communication processes. e-Commerce and SD practices have facilitated the acceleration of the NPD process.

However, innovation capability factors can have different impacts on innovation performance if these factors are treated in isolation or in different parts of the organization without establishing the synergistic relationships between them. Therefore, the integration of "mainstream" and "newstream" activities is a critical imperative for the creation of innovation-driven organizations. The mainstream provides the revenue for new products to be developed and commercialized in the newstream. Furthermore, a balance between "hard" (enablers) and "soft" (innovation values) are necessary for innovation to be successful and sustainable.

11.16 Implications for Managers

The results from the multiple cross-case analysis provide a theoretical and practical understanding for managers on the complex relationships between innovation capability, e-Commerce, sustainable development and new product development. This knowledge can assist managers to make more effective decisions in resource allocation for the creation of innovation-driven companies.

The research results can be used to justify the efficacy of innovation so that the present manager's perception of innovation as a technically-driven strategy is expanded to include innovation as a competitive business strategy. This book comprehensively informs managers about what factors they should apply incentives to, in terms of "what works, why and how it works" in creating innovative organizations. Based on the MCCA, several implications for innovation practice and research can be derived. There is a wider array of factors that can drive innovation capability and, thus, innovation. Accordingly, managers need to cross functional, geographical and business unit boundaries and identify additional sets of driving factors.

The MCCA indicates that innovation capability factors can have different impacts on innovation performance individually that will not materialize if these factors are treated in isolation or in different parts of the organization without establishing the synergistic relationships between these factors. For example, SDO does have a significant positive effect on innovation performance, especially when linked to the NPD process. All case study companies confirmed that a strategic configuration of sustainability considerations can bolster a firm's competitive position. The analysis demonstrates that business strategy is a major determinant for leveraging the company's innovation capabilities.

Review Questions

(1) What constitutes innovation capability in organizations? What are the key characteristics of innovative organizations?

(2) With respect to the five generations of innovation listed in Chapter 1, determine the category (1G to 5G) for each of the eight case studies. Support your answer with evidence from the case study.

(3) With reference to question 2, articulate key recommendations to a senior manager on what works, why it works and how it works in innovation-driven organizations.

References

Acs, Z.J. and Audretsch, D.B. (1988). Innovation in large and small firms: An empirical analysis. *American Economic Review*, 78, 678–690.

Afuah, A. (1998). *Innovation Management: Strategies, Implementation and Profits*. Oxford University Press, New York.

Ahmed, P.K. (1998). Benchmarking innovation best practice. *Benchmarking for Quality Management & Technology*, 5 (1), 45–58.

Allee, V. (2000). The value evolution: Addressing larger implications of an intellectual capital and intangibles perspective. *Journal of Intellectual Capital*, 1 (1), 17–32.

Amabile, T.M., Conti, R., Coon, H., Lazenby, J. and Herron, M. (1996). Assessing the work environment for creativity. *Academy of Management Journal*, 39 (5), 1154–1184.

Anand, J. and Singh, H. (1997). Asset redeployment, acquisitions and corporate strategy in declining industries. *Strategic Management Journal*, 18 (Summer Special Issue), 99–118.

Angle, H.L. (1989). Psychology and organizational innovation. In H. Van de Ven, H.L. Angle and M.S. Poole (Eds.), *Research on the Management of Innovation* (pp. 135–170). Ballinger/Harper & Row, New York.

Argote, L. and Ingram, P. (2000). Knowledge transfer: A basis for competitive advantage in firms. *Organizational Behavior and Human Decision Processes*, 82 (1), 150–169.

Atuahene-Gima, K. (1996). Differential potency of factors affecting innovation performance in manufacturing and services firms in Australia. *Journal of Product Innovation Management*, 13, 35–52.

Audretsch, D.B. (1995). Innovation, growth and survival. *International Journal of Industrial Organization*, 13, 441–457.

Azzone, G. and Noci, G. (1998). Seeing ecology and "green" innovations as a source of change. *Journal of Organizational Change Management*, 11, 94–111.

Balan, S. (1994). Manufacturing technology management: Key issues in the adoption, implementation and evaluation of advanced manufacturing technology. Ph.D. thesis, University of Canterbury, Christchurch, New Zealand.

Bansal, P. and Roth, K. (2000). Why companies go green: A model of ecological responsiveness. *Academy of Management Journal*, 43 (4), 717–748.

Baptista, R. and Swann, P. (1998). Do firms in clusters innovate more? *Research Policy*, 27, 525–540.

Barker, V.L. and Mueller, G.C. (2002). CEO characteristics and firm R&D spending. *Management Science*, 48, 782–801.

Barney, J.B. (1991). Firm resources and sustained competitive advantage. *Journal of Management*, 17, 99–120.

Baum, J.A.C., Calabrese, T. and Silverman, B.S. (2000). Don't go it alone: Alliance network composition and startups' performance in Canadian biotechnology. *Strategic Management Journal*, 21, 267–294.

Bessant, J., Francis, D., Meredith, S., Kaplinsky, R. and Brown, S. (2001). Developing manufacturing agility in SMEs. *International Journal of Technology Management*, 16 (4), 427–439.

Boer, H., Caffyn, S., Corso, M., Coughlan, P., Gieskes, J., Magnusson, M., Pavesi, S. and Ronchi, S. (2001). Knowledge and continuous innovation. *International Journal of Operations & Production Management*, 21, 490–503.

Bouty, I. (2000). Interpersonal and interaction influences on informal resource exchange between R&D researchers across organizational boundaries. *Academy of Management Journal*, 43, 50–65.

Brooker Group (2001). Technological innovation of industrial enterprise in Thailand. Project Synthesis prepared by The Brooker Group plc, for the regional workshops on Innovation in the Manufacturing Sector, 18–20 July, Bangkok and Penang.

Brown, S. and Bessant, J. (2003). The manufacturing strategy-capabilities links in mass customisation and agile manufacturing — An exploratory study. *International Journal of Operations & Production Management*, 23 (7), 707–730.

Burgelman, R.A. and Maidique, M.A. (1988). *Strategic Management of Technology and Innovation*. Irwin, Homewood, Illinois.

Caldeira, M.M. and Ward, J.M. (2003). Using resource-based theory to interpret the successful adoption and use of information systems and technology in manufacturing small and medium-sized enterprises. *European Journal of Information Systems*, 12, 127–141.

Cappelli, P. (2000). A market-driven approach to retaining talent. *Harvard Business Review*, 78 (1), 103–111.

Cerin, P. and Karlson, L. (2002). Business incentives for sustainability: A property rights approach. *Ecological Economics*, 40, 13–22.

Chan, C. and Swatman, P.M.C. (2000). From EDI to internet commerce: The BHP Steel experience. *Internet Research*, 10 (1), 72–83.

Chang, K., Jackson, J. and Grover, V. (2002). E-commerce and corporate strategy: An executive perspective. *Information & Management*, 40 (7), 663–675.

Chapman, P., James-Moore, M., Szczygiel, M. and Thompson, D. (2000). Building internet capabilities in SMEs. *Logistics Information Management*, 13 (6), 353–361.

Chesbrough, H. (2003). *Open Innovation*. Harvard Business School Press, Boston.

Clark, K.B. and Fujimoto, T. (1991). *Product Development Performance: Strategy, Organisation, and Management in the World Auto Industry.* Harvard Business School Press, Boston.

Cohen, J. and Levinthal, D.A. (1990). Absorptive capacity: A new perspective on learning and innovation. *Administrative Science Quarterly*, 35 (1), 554–571.

Considine, T.J., Jablonowski, C., Posner, B. and Bishop, C.H. (2004). The value of hurricane forecasts to oil and gas producers in the Gulf of Mexico. *Journal of Applied Meteorology*, 43 (9), 1270–1281.

Cooper, R.G. (1985). Industrial firms' new product strategies. *Journal of Business Research*, 13 (2), 107–121.

Cooper, R.G. and Kleinschmidt, E.J. (1996). Winning businesses in new product development. *Research-Technology Management*, 39 (4), 18–29.

Cooper, R.G., Edgett, S.J. and Kleinschmidt, E.J. (1998). Best practices for managing R&D portfolios. *Research Technology Management*, 41 (4), 20–33.

Coriat, B., Orsi, F. and Weinstein, O. (2003). Does biotech reflect a new science-based innovation regime? *Industry and Innovation*, 10 (3), 231–253.

Cormier, D., Magnan, M. and Morard, B. (1993). The impact of corporate pollution on market valuation. *Ecological Economics*, 8, 135–155.

Culkin, N. and Smith, D. (2000). An emotional business: A guide to understanding the motivations of small business decision takers. *Qualitative Market Research: An International Journal*, 3 (3), 145–157.

Damanpour, F. (1991). Organizational innovation: A meta-analysis of effects of determinants and moderators. *Academy of Management Journal*, 34, 555–590.

Damanpour, F. (1992). Organizational size and innovation. *Organization Studies*, 13, 375–402.

Debreceny, R., Putterill, M., Tung, L. and Gilbert, A.L. (2002). New tools for the determination of e-commerce inhibitors. *Decision Support Systems*, 34, 177–195.

Delaplace, M. and Kabouya, H. (2001). Some considerations about interactions between regulation and technological innovation. *European Journal of Innovation Management*, 4 (4), 179–185.

Derwall, J., Guenster, N., Bauer, R. and Koedijk, K. (2005). The eco-efficiency premium puzzle. *Financial Analysts Journal*, 61 (2), 51–64.

Dougherty, D. and Bowman, E.H. (1996). The effects of organizational downsizing on product innovation. *Journal of Product Innovation Management*, 13 (2), 173–174.

Dougherty, D. and Hardy, C. (1996). Sustained product innovation in large, mature organizations: Overcoming innovation-to-organisation problems. *Academy of Management Journal*, 39 (5), 1120–1153.

Dunphy, D., Griffiths, A. and Benn, S. (2003). *Organizational Change for Corporate Sustainability*. Routledge, New York.

Eisenhardt, K.M. (1989). Making fast strategic decisions in high-velocity environments. *Academy of Management Journal*, 32 (3), 543–576.

Eisenhardt, K.M. and Martin, J.A. (2000). Dynamic capabilities: What are they? *Strategic Management Journal*, 21, 1105–1121.

Firth, R.W. and Narayanan, V.K. (1996). New product strategies of large, dominant product manufacturing firms: An exploratory analysis. *Journal of Product Innovation Management*, 13, 334–347.

Fuchs, P.H., Mifflin, K.E., Miller, D. and Whitney, J.O. (2000). Strategic integration: Competing in the age of capabilities. *California Management Review*, 42, 118–147.

Fulk, J. and DeSanctis, G. (1995). Electronic communication and changing organizational forms. *Organization Science*, 6 (4), 337–350.

Galunic, D.C. and Rodan, S. (1998). Resource combinations in the firm: Knowledge structures and the potential for Schumpeterian innovation. *Strategic Management Journal*, 19, 1193–1201.

Gebauer, J. and Scharl, A. (2003). Between flexibility and automation. In C. Steinfield (Ed.), *New Directions in Research on E-Commerce* (pp. 147–169). Purdue University Press, Indiana.

Gertakis, J. (2001). Maximising environmental quality through EcoReDesign. In M. Charter and U. Tichner (Eds.), *Sustainable Solutions: Developing Products for the Future*. Greenleaf Publishing, Sheffield.

Gittelman, M. and Kogut, B. (2003). Does good science lead to valuable knowledge? Biotechnology firms and the evolutionary logic of citation patterns. *Management Science*, 49 (4), 366–382.

Gleadle, P. (1999). The interface between finance and new product development encouraging a climate of innovation. *Management Accounting*, 77 (7), 24–25.

Gloet, M. and Terziovski, M. (2002). The relationship between knowledge management and innovation performance: A qualitative analysis. Paper presented at the 7th International Conference on ISO 9000 and TQM (7-ICT), Melbourne, Hong Kong Baptist University.

Goldsmith, S. and Samson, D. (2002). Sustainable development — State of the art asking the questions. Issues paper prepared for the Australian Business Foundation, The University of Melbourne, Australia.

Grant, R.M. (1996). Toward a knowledge-based theory of the firm. *Strategic Management Journal*, 17 (Winter Special Issue), 109–122.

Greer, Jr., W.R. and Liao, S.S. (1986). An analysis of risk and return in the defense market: Its impact on weapon system competition. *Management Science*, 32, 1259–1273.

Gulati, R. and Garino, J. (2000). Get the right mix of bricks and clicks. *Harvard Business Review*, 78 (3), 107–114.

Hagedoorn, J. and Cloodt, M. (2003). Measuring innovative performance: Is there an advantage in using multiple indicators? *Research Policy*, 32, 1365–1379.

Hall, P.A. and Soskice, D.W. (2001). *Varieties of Capitalism: The Institutional Foundations of Comparative Advantage*. Oxford University Press.

Hansen, M.T., Chesbrough, H.W., Nohria, N. and Sull, D.N. (2000). Networked incubators: Hothouses of the new economy. *Harvard Business Review*, 78 (5), 74–88.

Haque, B. and James-Moore, M. (2005). Performance measurement experiences in aerospace product development processes. *International Journal of Business Performance Management*, 7, 100–122.

Harrington, H.J. (1995). Continuous versus breakthrough improvement: Finding the right answer. *Business Process Reengineering & Management Journal*, 1 (3), 31–49.

Hart, S.L. (1997). Beyond greening: Strategies for a sustainable world. *Harvard Business Review*, 75 (1), 66–77.

Haynes, P.J., Becherer, R.C. and Helms, M.M. (1998). Small and mid-sized businesses and internet use: Unrealized potential? *Internet Research: Electronic Networking Applications and Policy*, 8 (3), 229–235.

Hill, C.W.L. and Rothaermel, F.T. (2003). The performance of incumbent firms in the face of radical technological innovation. *Academy of Management Review*, 28, 257–274.

Holak, S.L., Parry, M.E. and Song, X.M. (1991). The relationship of R&D/sales to firm performance: An investigation of marketing contingencies. *Journal of Product Innovation Management*, 8, 267–282.

Hunt, C.B. and Auster, E.R. (1990). Proactive environmental management: Avoiding the toxic trap. *Sloan Management Review*, 31 (2), 7–18.

Intarakumnerd, P., Chairatana, P.A. and Tangchitpiboon, T. (2002). National innovation system in less successful developing countries: The case of Thailand. *Research Policy*, 31 (8–9), 1445–1457.

Jha, S., Noori, H. and Michela, J.L. (1996). The dynamics of continuous improvement: Aligning organisational attributes and activities for quality and productivity. *International Journal of Quality Science*, 1 (1), 19–47.

Jurkus, A.F. (1990). Requiem for a lightweight: The Northrop F-20 strategic initiative. *Strategic Management Journal*, 11, 59–68.

Kanter, J. (1999). Knowledge management, practically speaking. *Information Systems Management*, 16 (4), 7–15.

Kanter, R.M. (1984). *The Change Masters: Innovation and Entrepreneurship in the American Corporation*. Simon & Schuster, New York.

Kanter, R.M. (1989). Swimming in newstreams: Mastering innovation dilemmas. *California Management Review*, 31, 45–69.

Kiernan, M.J. (1996). Get innovative or get dead. *Business Quarterly*, Autumn, 51–58.

Kim, W.C. and Mauborgne, R. (1999). Strategy, value innovation, and the knowledge economy. *Sloan Management Review*, 40 (3), 41–54.

Kim, W.C. and Mauborgne, R. (2004). Blue ocean strategy. *Harvard Business Review*, 82 (10), 76–84.

Klassen, R. and McLaughlin, C. (1996). The impact of environmental management on firm performance. *Management Science*, 42 (8), 1199–1214.

Konings, J. and Roodhooft, F. (2002). The effect of e-business on corporate performance: Firm level evidence for Belgium. *The Economist*, 150, 569–581.

Koput, K.W., Smith-Doerr, L. and Powell, W.W. (1996). Interorganizational collaboration and the locus of innovation. *Administrative Science Quarterly*, 41 (1), 116–145.

Laursen, K. and Foss, N.J. (2003). New human resource management practices, complementarities and the impact on innovation performance. *Cambridge Journal of Economics*, 27, 243–263.

Lawson, B. and Samson, D. (2001). Developing innovation capability in organizations: A dynamic capabilities approach. *International Journal of Innovation Management*, 5 (3), 377–400.

Lawson, B. and Samson, D. (2003). Implementing e-business change: An innovation management perspective. In D. Samson (Ed.), *E-Business: Value Creation for Management*. McGraw-Hill, Sydney.

Longenecker, C.O. and Meade, II, W.K. (1995). Marketing as a management style. *Business Horizons*, 38 (4), 77–83.

Mabert, V.A., Muth, J.F. and Schmenner, R.W. (1992). Collapsing new product development times: Six case studies. *Journal of Product Innovation Management*, 9, 200–212.

McCutcheon, D.M. and Meredith, J.R. (1993). Conducting case study research in operations management. *Journal of Operations Management*, 11, 239–256.

McGourty, J., Tarshis, L.A. and Dominick, P. (1996). Managing innovation: Lessons from world class organisations. *International Journal of Technology Management*, 11, 354–368.

Metz, I., Terziovski, M. and Samson, D. (2004). Future research agenda based on an integrated innovation capability model (IICM). Unpublished paper, The University of Melbourne.

Miles, I. (2000). Service innovation: Coming of age in the knowledge based economy. *International Journal of Innovation Management*, 4 (4), 371–389.

Miles, M.B. and Huberman, A.M. (1994). *Qualitative Data Analysis* (2nd ed.). Sage, Thousand Oaks, CA.

Miles, R.E. and Snow, C.C. (1992). *Causes of Failure in Network Organizations*. Free Press, New York.

Morss, R.E. and Hooke, W.H. (2005). The outlook for U.S. meteorological research in a commercializing world: Fair early, but clouds moving in? *Bulletin of the American Meteorological Society*, 86 (7), 921–936.

Mort, J. and Knapp, J. (1999). Integrating workspace design, web-based tools and organizational behavior. *Industrial Research Institute*, 42 (2), 33–40.

Ngai, E.W.T. and Wat, F.K.T. (2002). A literature review and classification of electronic commerce research. *Information & Management*, 39, 415–429.

Noori, H. and Chen, C. (2003). Applying scenario-driven strategy to integrate environmental management and product design. *Production and Operations Management*, 12 (3), 353–368.

Numprasertchai, S. and Igel, B. (2004). Indicators for measuring university research performance in Thailand. In *Proceeding of the International Symposium on Management of Technology (ISMOT'04)* (pp. 479–483). Hangzhou, China.

OECD (2000). Science, technology and innovation in the new economy. *Policy Brief,* OECD Observer, September, 1.

Omta, S.W.F. (1995). *Critical Success Factors in Biomedical Research and Pharmaceutical Innovation.* Kluwer, Dordrecht.

Omta, S.W.F., Bouter, L.M. and van Engelen, J.M.L. (1997). Management control of biomedical research and pharmaceutical innovation. *Technovation,* 17 (4), 167–179.

Ottman, J.A.R. and Reilly, W.K. (1998). *Green Marketing: Opportunity for Innovation.* NTC Business Books, Lincolnwood.

Pawar, K.S. and Sharifi, S. (1997). Physical or virtual team collocation: Does it matter? *International Journal of Production Economics,* 52 (3), 283–290.

Phan, D.D. (2002). E-business development for competitive advantage: A case study. *Information & Management,* 40, 581–590.

Polonsky, M.J. (2001). Re-evaluating green marketing: A strategic approach. *Business Horizons,* 44 (5), 21–31.

Porter, M.E. (2001). Strategy and the internet. *Harvard Business Review,* 79 (3), 63–78.

Porter, M.E. and Stern, S. (1999). *The New Challenge to America's Prosperity: Findings from the Innovation Index.* Council on Competitiveness, Washington.

Porter, M.E. and van der Linde, C. (1995). Toward a new conception of the environment-competitiveness relationship. *Journal of Economic Perspectives,* 9 (4), 97–118.

Powell, W.W., Koput, K.W. and Smith-Doerr, L. (1996). Interorganizational collaboration and the locus of innovation: Networks of learning in biotechnology. *Administrative Science Quarterly,* 41, 116–145.

Prahalad, C. and Hamel, G. (1990). The core competencies of the corporation. *Harvard Business Review,* 68 (3), 79–81.

Prybutok, V.R. and Ramasesh, R. (2005). An action-research based instrument for monitoring continuous quality improvement. *European Journal of Operational Research,* 166, 293–309.

Rennings, K. (2000). Redefining innovation/eco-innovation research and the contribution from ecological economics. *Ecological Economics,* 32, 319–332.

Roberts, L. (1996). Improving the environmental performance of firms: The experience of two metal working companies. *Journal of Cleaner Production*, 4 (3–4), 175–187.

Rodan, S. and Galunic, C. (2004). Knowledge heterogeneity in managerial networks and its effect on individual performance. *Strategic Management Journal*, 25 (6), 541–562.

Roffe, I. (1999). Innovation and creativity in organisations: A review of the implications for training and development. *Journal of European Industrial Training*, 23 (4–5), 224–237.

Rothwell, R. (1994). Towards fifth-generation process innovation. *International Marketing Review*, 11 (1), 7–31.

Rzakhanov, Z. (2004). Innovation, product development and market value: Evidence from the biotechnology industry. *Economics of Innovation and New Technology*, 13 (8), 747–760.

Salazar, A., Hackney, R. and Howells, J. (2003). The strategic impact of internet technology in biotechnology and pharmaceutical firms: Insights from a knowledge management perspective. *Information Technology and Management*, 4 (2–3), 289–301.

Salomo, S., Steinhoff, F. and Trommsdorff, V. (2003). Customer orientation in innovation projects and new product development success — The moderating effect of product innovativeness. *International Journal of Technology Management*, 26, 442–463.

Samson, D. (2003). *e-business*. McGraw-Hill, Australia.

Samson, D. and Daft, R. (2003). *Management*. Thompson, Australia.

Samson, D. and Terziovski, M. (1999). The relationship between total quality management practices and operational performance. *Journal of Operations Management*, 17 (4), 393–409. [*Won 1999 Journal of Operations Management Best Paper Award*]

Schroeder, R.G., Bates, K.A. and Junttila, M.A. (2002). A resource-based view of manufacturing strategy and the relationship to manufacturing performance. *Strategic Management Journal*, 23, 105–117.

Shafer, S.M., Smith, H.J. and Linder, J.C. (2005). The power of business models. *Business Horizons*, 48, 199–207.

Sharfman, M.P., Meo, M. and Ellington, R.T. (2000). Regulation, business, and sustainable development: The antecedents of environmentally conscious technological innovation. *The American Behavioural Scientist*, 44, 277–302.

Sharfman, M.P., Shaft, T.M. and Tihanyi, L. (2004). A model of the global and institutional antecedents of high-level corporate environmental performance. *Business & Society*, 43 (1), 6–36.

Shrivastava, P. (1995). Environmental technologies and competitive advantage. *Strategic Management Journal*, 16, 183–200.

Soderquist, K. (1996). Managing innovation in SMEs: A comparison of companies in the UK, France, and Portugal. *International Journal of Technology Management*, 12 (3), 291–305.

Sohal, A., Terziovski, M. and Zutshi, A. (2003). Team-based strategy at Varian Australia: A case study. *Technovation*, 23, 349–357.

Souder, W.E., Song, X.M. and Kawamura, K. (1998). America's edge in new product R&D. *Research Technology Management*, 41 (2), 49–56.

Subramanian, A. and Nilakanta, S. (1996). Organizational innovativeness: The relationship between organizational determinants of innovation, types of innovations, and measures of organizational performance. *International Journal of Management Science*, 24, 631–647.

Tam, P.W. (2003). Cloud over Sun Microsystems: Plummeting computer prices. *Wall Street Journal*, 16 October, pp. A1–A16.

Teece, D.J. and Pisano, G.P. (1994). The dynamic capabilities of firms: An introduction. International Institute for Applied Systems Analysis (IIASA) Working Paper WP-94-103.

Teece, D.J., Pisano, G. and Shuen, A. (1997). Dynamic capabilities and strategic management. *Strategic Management Journal*, 18, 509–533.

Terziovski, M. (2001). The effects of continuous improvement and innovation management practice on small to medium enterprise (SME) performance. Paper presented at the Academy of Management Conference (Operations Division), Washington, DC.

Terziovski, M. (2002). Achieving performance excellence through an integrated strategy of radical innovation and continuous improvement. *Measuring Business Excellence*, 6 (2), 5–14.

Terziovski, M. (2003). The relationship between networking practices and business excellence. *Measuring Business Excellence*, 7 (2), 78–92.

Terziovski, M., Sohal, A. *et al.* (2002). Best practice in product innovation at Varian Australia. *Technovation*, 22 (9), 561–569.

Theyel, G. (2000). Management practices for environmental innovation and performance. *International Journal of Operations and Production Management*, 20, 249–266.

Tsai, W. and Ghoshal, S. (1998). Social capital and value creation: The role of intrafirm networks. *Academy of Management Journal*, 41 (4), 464–476.

von Hippel, E. (1988). *The Sources of Innovation*. Oxford University Press, New York.

von Hippel, E., Thomke, S. and Sonnack, M. (1999). Creating break-throughs at 3M. *Harvard Business Review*, 77 (5), 47–57.

Walls, L. and Quigley, J. (1999). Learning to improve during system development. *European Journal of Operational Research*, 119, 495–509.

Wernerfelt, B. (1984). A resource-based view of the firm. *Strategic Management Journal*, 5, 171–180.

Wheelen, T.L. and Hunger, J.D. (2002). *Strategic Management and Business Policy*. Pearson Prentice Hall, New Jersey.

Wheelen, T.L. and Hunger, J.D. (2004). *Concepts in Strategic Management and Business Policy*. Prentice Hall.

Wolfe, R.A. (1994). Organizational innovation: Review, critique and suggested research directions. *Journal of Management Studies*, 31, 405–431.

Wright, P.M. and Snell, S.A. (1998). Toward a unifying framework for exploring fit and flexibility in strategic human resource management. *Academy of Management Review*, 23 (4), 756–772.

Yin, R.K. (1989). *Case Study Research: Design and Methods* (Revised Edition). Sage, London.

Yin, R.K. (2003). *Case Study Research: Design and Methods* (3rd ed.). Sage, Thousand Oaks, California.

Yip, G. and Dempster, A. (2005). Using the internet to enhance global strategy. *European Management Journal*, 23 (1), 1–13.

Zahra, S.A. and George, G. (2002). Absorptive capacity: A review, reconceptualization, and extension. *Academy of Management Review*, 27 (2), 185–203.

Index

absorptive capacity   58, 111, 118, 147, 154, 167, 200, 202, 219, 228, 231, 233

administrative intensity   27

adoption   1, 21, 23, 25–28, 32, 35, 36, 38, 39, 131, 150, 210, 226

alliances   10, 26, 58, 79, 81, 85, 137, 167, 202, 207, 209, 229

applied knowledge management   61

assets   8, 59, 60, 94, 95, 111, 122, 158, 162, 164, 173, 175, 192, 201, 206

automation   53, 95, 144, 201

autonomous business units   122, 211

best practice   6, 13, 16, 141, 157, 169, 192, 207

BHP Steel   39

biotechnology   vii, 10, 11, 14, 17, 51–53, 55, 58, 61, 63, 65, 157–159, 173, 174, 192, 200, 223

Blackmores   7

branding   100, 117, 220

Business Excellence   12, 37

business-to-consumer   39

buyer value   2, 192

Caterpillar   15, 87–99, 101, 103, 197, 207, 225

clusters   24, 25, 39

codified knowledge   17, 157, 215

competitive advantage   8, 14–17, 36, 41, 42, 51, 64, 65, 73, 77, 84, 88, 90, 91, 105–107, 117, 121, 137, 141, 152, 158, 159, 165, 167, 172, 174, 192, 196, 223, 225

competitor-based benchmarking   85

compulsory license   171

conceptual framework   118

consumer trust   39

continuous customer proximity   54

continuous feedback   82, 201, 204, 228

continuous improvement   2, 27, 61, 79, 92, 100, 103, 118, 132, 135, 138, 145, 146, 205–207

continuous learning   91, 103

copyright   22, 164, 167

core competency   129, 162

core values   124, 130, 145, 214

corporate culture   15, 17, 60, 69, 107, 157, 166, 174, 215

corporate environmental responsiveness   149

corporate governance   124, 125, 129, 214, 224

corporate philosophy   55, 145, 148, 150, 152, 215, 216, 224

corporate skill base   146, 154, 202, 219, 231

corporate strategy   39, 62, 107,
    124, 129, 162, 222, 224
cost advantage   99, 116, 117
cost reduction   4, 97, 225
creative destruction   26
cross-unit development   152, 221
customer feedback   79, 80, 102,
    197
customer involvement   15, 74,
    80, 85, 151, 152, 205, 221
customer orientation   5, 65, 69, 74,
    80, 83, 85, 101, 104, 145, 146,
    151, 191, 199, 202, 204–206,
    220, 221, 228, 229, 232
customer satisfaction   27, 69, 79,
    83, 92, 100

data collection   173, 227
dendrimer science   157, 159, 160,
    162, 164, 167, 170, 202
differentiation advantage   99,
    117, 144, 198
diffusion   60, 61, 70, 150, 151,
    153, 178, 193, 207, 208, 211,
    212, 226, 230
Dishlex   7
DPEC   16, 121–134, 137–139,
    194, 198, 202, 208, 211, 214,
    217, 221, 224, 226
dynamic capabilities   19, 73, 141
dynamic organizational capabilities
    51

e-Commerce   vii, 4, 6–9, 14, 16,
    18, 20–23, 26, 27, 30, 32,
    36–40, 43, 44, 46–49, 52, 64,
    97–99, 104, 105, 109,
    114–116, 118, 119, 121,

124–126, 129, 130, 141, 142,
    147, 150–152, 154, 191, 194,
    195, 217, 223–227, 232–234
e-Communication   130, 131,
    138, 139, 221
Electronic Data Interchange   38
employee participation   83
enabling factors   4, 17, 20
entrepreneurial networks   30,
    31
entrepreneurship   xi, xii, 102,
    105, 179, 197, 214, 215
environmental innovations   42,
    64, 223
environmental regulation   22, 90,
    96, 109, 148, 222, 223, 232
Eurocopter   15, 73–77, 79–86,
    193, 196, 200, 201, 204, 205,
    207, 210, 213, 214, 216, 220,
    223
exchange networks   62, 70
external communication   30, 42,
    47
external partners   15, 47, 58–60,
    70, 74, 81, 94, 193, 200, 203,
    228, 233
extrinsic motivation   35, 69, 113,
    196

feedback   3, 65, 67, 79, 80, 82,
    101, 102, 186, 197, 201, 204,
    228
fifth generation innovation   4
financing   90, 203
forecasting   143
fourth generation innovation   3, 4
free-riding   192
functional differentiation   27, 28

General Electric (GE) 10
global competitiveness 121
goal orientation 145, 215
Google 130
government regulation 22, 23,
  42, 47, 48, 95, 109, 113, 127,
  129, 148, 205, 224

heterogeneity of knowledge 141,
  144, 198
heterogeneous knowledge 84,
  153, 196
hierarchy 27
horizontal approach 85, 205,
  210, 212, 233
HRM systems 45
human capital 164

ideas management 102, 197,
  211, 212, 230, 233
incremental research 146
individual knowledge 62, 68, 84,
  151, 196, 208, 209, 211, 213,
  229
innovation champions 69, 196,
  213
innovation culture (3M corporation)
  27, 41
innovation driver 4, 6, 7, 13, 15,
  16, 18, 105, 118, 141, 142,
  147, 191, 204, 206, 207, 209,
  222, 233
innovation enabler 15, 16, 87,
  104, 105
innovation implementation 35
innovation points 127
innovation practices 16, 18, 142,
  191, 208, 209

innovation rate 54, 57, 59, 71, 210
intangible asset 8
integrated innovation capability
  model 14, 19, 37, 118, 126,
  191
intellectual capital 7, 8, 33, 126,
  188, 192
intellectual property 17, 123, 126,
  132, 157–160, 162, 164–166,
  168–175, 192, 193, 199,
  203, 215, 217–219, 221,
  231, 232
intellectual property strategy 17,
  157, 164, 165, 174, 221
internet 8, 23, 33, 36, 38, 64,
  65, 98, 99, 101, 106, 107, 114,
  130, 133, 138, 173, 178, 185,
  186, 202–204, 208, 210, 211,
  225, 226
internet-based technology 97,
  98, 114, 130, 131, 150, 202,
  211, 212, 225, 226, 230
intra-company knowledge diffusion
  70, 193
intrinsic motivation 69
Invincible Company 17, 177, 179,
  189, 194, 199, 205, 208, 215
ISO 14000 148
ISO 14001 62, 63, 149, 222,
  224
ISO 9000 100, 147

Japan 143,
job rotation 29, 84, 137, 153, 196
joint ventures 94, 146, 202

Kambrook 7
knowledge creation 78, 220

knowledge exchange   68, 81, 84,
    150, 196, 199, 213, 215, 226,
    228, 230
knowledge management (KM)
    138, 194
knowledge storage   150, 226
knowledge transfer   10, 16, 65,
    70, 139, 150, 208, 209, 221,
    225, 226, 229
knowledge-based centers   112

lead user   36
leadership   5, 19, 31–33, 43,
    48, 91, 94, 96, 101, 103,
    104, 106, 109, 125, 194,
    195, 227
leadership capabilities   19
leadership skills   32
licensing   137, 158–160, 164,
    167, 172
local knowledge   146, 200, 206
logistics systems   89

mainstream   13, 15, 18, 30, 73,
    74, 77–79, 82, 83, 84, 85,
    86, 93, 193, 195, 220,
    227, 234
market entry   102
market knowledge   5, 25, 36, 37,
    49, 109
marketable research outcomes
    153, 217, 219, 231
mission statement   54, 55,
    101, 108, 124, 142, 145,
    155, 216
multiple cross-case analysis
    (MCCA)   13, 18, 191, 199,
    233, 234

networked incubators   12
New Product Development (NPD)
    78, 99, 104, 115, 128, 141,
    151, 163, 219
newstream   13, 15, 18, 73, 74,
    77–79, 81–86, 93, 101, 193,
    195, 220, 227, 234
non-commercialized innovations
    56
non-profit organizations   25, 26

old technologies   57, 66
operational excellence   56, 64,
    223
organizational knowledge   33, 43,
    52
organizational structure   4, 5, 12,
    27, 28, 47, 57, 58, 102, 104,
    137, 138, 141, 145, 153, 154,
    163, 173, 182–184, 189, 191,
    194, 195, 205, 206, 210,
    211–213, 227, 229, 230

partial team integration   127
partnerships   4, 10, 11, 22,
    24, 45–47, 92, 109, 143,
    167, 202
performance advantage   108
performance measures   9, 219
Pooled Development Fund (PDF)
    160
process innovation   13, 15, 42,
    74, 78, 85, 86, 116, 133–135,
    138, 204,
    210, 220
product chain   148
product conceptualization phase
    222

product development cycles 121

product development process 10, 14, 20, 70, 80, 85, 164, 168, 173, 174, 193, 205, 206, 228

product distribution structure 145

product orientation 14, 51, 55, 204, 216, 218, 231, 234

productivity 26, 27, 34, 39, 92, 93, 98, 100, 101, 106

professional knowledge 33, 34, 207, 209, 229, 233

public funding 170

QFD 131

radical customer orientation 85

radical improvement 27

recoverable manufacturing 40, 95

relationship management 107, 173

research phase 95

restructure 7, 10, 11, 146, 233

reverse engineering 170

risk 9, 29, 30, 31, 41, 47, 57, 60, 65, 69, 97, 102, 110, 112, 158, 165, 166, 169, 174, 193, 202, 214, 215, 218, 219, 230, 231, 232

risk-taking 102, 214, 215, 231

sequential vertical structure 85, 205, 210, 212, 230, 233

simultaneous innovation processes 59, 82, 200, 203

Six Sigma 88, 93, 94, 99, 101, 102, 110, 111, 113, 118, 201, 214, 220

slack resources 29, 109, 118

SME 8, 11, 12, 29, 33, 37, 79, 80, 100, 102, 132, 159, 178, 203

solution-focused approaches 133, 198

staff performance 184, 199, 228

Starpharma 17, 157, 159, 160–175, 192, 198, 199, 202, 203, 206, 208, 210–212, 215, 217, 221, 224, 227

strategic alliances 10, 79, 81, 137

strategic management 1, 26, 41

strategic partnering 11

strategic patenting 159

strategic research 146

Sun Microsystems (SMS) 15, 41, 74, 105–119, 194, 197, 198, 201, 202, 205, 208, 211, 214, 217, 220, 223, 224, 226

supplier protocols 40

sustainable development 4, 6, 7, 14, 16, 18, 20–23, 26, 27, 30, 32, 36, 37, 40, 41, 43, 44, 46–49, 52, 62–64, 77, 90, 95–97, 99, 104, 105, 112, 113, 115, 116, 118, 119, 129, 139, 141, 143, 145, 147–149, 151, 153–155, 179, 191, 197, 205, 214, 221–224, 226, 232, 234

sustained customer value 83, 85, 205, 206, 228

synergistic innovation capability 14

systematic customer integration 79, 83, 196

systematic job rotation 84, 196

teamwork   9, 32, 34, 85, 91, 103, 107, 124, 128, 195, 234

technical knowledge   33, 182, 207, 218

technological synergy   36

technology change   23, 35

technology innovation   2, 29, 35, 192, 227

third generation innovation   3, 4

time to market   165, 172

Total Quality Management (TQM)   45, 131, 147

total solution management   55, 56, 204, 216, 218, 231, 234

trade secret   164

trademarks   123, 164–167, 172, 175

triangulated methodology   43

UGM   15, 87–95, 97–104, 194, 197, 201, 205, 207, 211, 214, 216, 220, 223, 225

umbrella arrangement   167

Vaisala   16, 143–155, 194, 198, 202, 205, 208, 211, 214, 217, 221, 224, 226

value chain   27, 94, 174

value creation   73

value innovation   2, 54, 57, 70, 71, 81, 90, 146, 192, 193, 217, 227, 234

value innovation strategy   57, 146

vertical differentiation   29

virtual corporations   10

virtual organization   113, 173, 205

waste minimization   40, 77, 95, 223`

"we do better" strategy   17, 182, 189, 194, 212